Globalisation
and
Information Technology

Globalisation and Information Technology

By

Dr. M. Lakshmi Narasaiah
M.A., Ph.D.

Professor of Economics,
Coordinator, Department of M.B.A.
Sri Krishnadevaraya University Post-graduate Centre,
Kurnool–518 002
Andhra Pradesh (India)

DISCOVERY PUBLISHING HOUSE
NEW DELHI

First Published–2005

ISBN: 81-7141-947-X

Published by:

DISCOVERY PUBLISHING HOUSE

4831/24, Prahlad Street, Ansari Road, Darya Ganj
New Delhi–110 002 (India)
Phone: 23279245, • Fax: 91-11-23253475
e-mail: dphtemp@indiatimes.com

Printed at:

Amit Enterprises, Delhi

Preface

Galloping advances in information technology promises to give us instant access to all worlds' knowledge. But how will human memory fare against the rise of the super-machine? If the architects of technology's next great leap forward are to be believed, all knowledge may soon be shrunk to vanishing point. Nanotechnology, or computing carried out at the scale of atoms, is their by word for the future. With its awesome potential, scientists have recently argued, around 11 million 400-page volumes could be stored and primed for instant viewing on device the size of a human palm.

The ink may still be fresh on these blueprints, but the elixir of portable omniscience no longer seems so far away. Seemingly cast-iron laws of ever increasing computer power, along with the rise of powerful new technologies, appear to point to horizon where all that can be known and remembered can be transferred to machines with which human beings then interact at will. And it is a future that for some is already spelling big trouble for the brain.

Surveys Point to Yawing Gaps in General Knowledge

Computers not only distract us from contemplation of deeper valleys; they discourage from contemplation itself. As surveys repeatedly show, knowledge of history, literature, geography and even current affairs seem to be on a steep decline: 60 per cent of adult Americans cannot recall the name of the President who ordered the dropping of the first atomic bomb, just as 77 per cent of young Britons are perplexed by words Magna Carta. The day of the nano-shrunk library could soon come, but will any of its users be able to remember a single line of poetry?

The connection between these yawning gaps in general knowledge and the information technology is by no means established, but a host of thinkers in different fields are sure the issues is one that will shortly become all to pertinent. At the same time as it help us and extends our physically capabilities, it diminishes our individual faculties. This is a vital question, one which has been around for a long time.

Good or evil, writing has nevertheless formed one of the main tools in the evolution of human memory. Indeed it is civilisation's unrelenting hunger for placing memory in external stores—cave paintings then manuscripts, libraries, printed works and finally computers - that has supported the entire march of the species. Each of these new technologies has helped humans "off-load" their memories. Pre-literate societies, for instance, depended on oral tradition for their expertise—a practice undermined by the flaws of over worked brains, though fertile ground for epic poetry. Through the written word, memories were freed from the head: knowledge could be stored for retrieval in books, and then redrafted into the sort of novel and complex codes on which modern society is founded.

Becoming God Memory Managers

The benefits of storing memory outside the brain are unquestionable, but the invention of printing over 500 years ago followed by the post-war onset of computing have added a new note to the process: that of thundering acceleration. One simple equation has come to embody this. It stipulates that computing power—defined in terms of capacity and speed per unit cost—doubles every two years. The trend has held for the last 40 years. Should it continue as expected to around 2020, a personal computer by that year will have exactly the same processing power as a single human brain. Add the promised marvels of nantechnology, optical and quantum computing, and machines might reach utterly daunting proportions One penny's worth of computing circa 2099 will have a billion times greater computing capacity than all humans on Earth.

For many cognitive scientists, relations between mind and machine are already undergoing drastic reconfiguration. "Distributed intelligence" is the new maxim encapsulating all systems in which individuals and computers mesh to carry out a collective task, whether it be landing an aircraft or tracking share prices. The Internet is so far the crowning glory—a system that in principle might combine individual users into a potent group min.

All of this may sound abstract, but the effects on memory are being felt now. Facts and figures no longer take pride of place in school curricula. Within the past two years, South Korea, Singapore and Hong Kong—havens of rote learning—have debated plans to axe huge swathes of standard classroom study, Experts in education stress that students must learn to be adaptable, skilled in manipulating symbols, able to respond to new situations; in short, ready to deal with the new economy, a realm where the computer is king.

Dr. M. Lakshmi Narasaiah

Contents

	Preface	*v*
1.	Globalisation: *A Moral Imperative*	1
2.	The Nation-State and Globalisation	4
3.	Urbanisation and Globalisation	8
4.	Globalisation and Knowledge Divide	11
5.	The Truth About Global Competition: *The Economic Myths Behind Globalisation*	14
6.	High World Trade Growth vs Output—WTO Sees Link to Globalisation	21
7.	Renewing the State	24
8.	Free Trade as Peacemaker: *The Benefits of an Open World Trading System*	26
9.	Democracy and the Market Economy	32
10.	Myths and Illusions	38
11.	Growing Complexity in International Economic Relations Demands Broadening and Deepening of the Multilateral Trading System	40
12.	World Trade—The Next Challenge	44
13.	What was Wrong with Structural Adjustment: *In Defence of a Much-Maligned Strategy*	49
14.	Add Value, go Global: *Can Southern Firms Break into Export Markets?*	55
15.	What's Driving Migration	59

16. Crisis and New Orientation of Development Policy 64

17. The End of the Old Order: *No Guarantee for Peace and Prosperity* 71

18. Employment and Promoting Ecology: *How a Service Culture Could Put People Back to Work* 74

19. Give Developing Countries A More Favorable Deal: *An Assessment of the World Trade Conference in Doha* 80

20. E Learning-Designing Tomorrow's Education 89

21. The dot.bomb Syndrome 91

22. The Electronic Gap 94

23. When Computers Chip Away at Our Memories 96

24. Fleeing the dot.com Era 100

25. Labour Pains: The Birth of a Movement 106

26. Shhh... they're Listening 109

27. Inclusion or Exclusion 113

28. Net Gains or Net Dream? 117

29. Wiring up the Ivory Towers 121

30. Information Technology Outsourcing Goes Global 125

Bibliography 131

Index 133

1

Globalisation: A Moral Imperative

Globalisation has become today's buzzword. It has also become a battle ground for two radically opposed groups. There are the anti-globalists, who fear globalisation and stress only its downside, seeking therefore powerful interventions aimed at taming, if not (unwittingly) crippling it. Then there are the "globalist" (a class to which I belong) who celebrate globalisation instead, emphasize its upside, while seeking only to ensure that its few rough edges be handled through appropriate policies that serve to make globalisation yet more attractive.

Many anti-globalists consider the central problem of globalisation to be its amorality, or even its immorality. But these critics have too blanket an approach to globalisation. The word covers a variety of phenomena that characterize an integrating world economy: trade, short-term capital flows, direct foreign investment, immigration, cultural convergence et al. The sins of one of the above cannot be visited upon the virtues of another. Some are benign even when largely unregulated whereas others can be fatal if left wholly to the marketplace.

In particular, the freeing of trade is largely benign: if I exchange some of my toothpaste for some of your toothbrushes, we will both be better off than if we did not trade at all. It would require a wild imagination, and a deranged mind, to think that such freeing of trade leads to debilitating economic crises. Equally, it is illogical to believe, as non-economists who fear globalisation do, that freeing of trade is bad because the freeing of short-term capital flows led to a debilitating financial and economic crises and could do so again. In fact, while there are some obvious simulates between free trade and free capital flows, e.g. that segmentation of markets creates efficiency losses, the economic and political dissimilarities are even more compelling and policy makers cannot ignore them.

Anti-globalist critics are in fact often reacting viscerally to a much larger issue: the victory of capitalism over its arch rival, communism, For campus idealists who have always looked for alternatives to what they conventionally consider to be the greed and lack of social conscience that characterize capitalism, the situation is psychologically intolerable Some have turned to street theatre, nihilism and the anti-intellectualism that has been manifest in the last few years. The more sophisticated have succumbed to a stereotypical representation of corporation as the evil forces of capitalism that have captured the state, democratic institutions, and even international bodies such as the World Trade Organisation.

What these critics often forget is that certain economic freedoms are basic to prosperity and social well-being under any conditions, and are thus of the highest moral value. Property right and markets, for instance, provide incentives to produce and allocate resources efficiently, and can in turn strengthen democracy by allowing a means of sustenance out side pervasive government structures. The quality and breadth of democracy can then be enlarged as excluded groups, such as women and the poor, are pulled into literacy, gainful employment and better health through higher public spending or the spread of economic incentives

Critics nevertheless go on to maintain that the global spread of free markets and free trade is responsible for continuing poverty in poor countries, and for alleged growth in inequality between and within countries. Labour unions in the rich countries also fear that trade in cheap labour-using goods from poor countries.

But I do not think these concerns are well-founded. In India which has almost a quarter of the world's poor, there is good evidence that autarchic and anti-market policies produced abysmally low growth rates at 3.5 per cent annually over a quarter of a century, with a correspondingly negligible impact on poverty has declined. Higher growth rates in turn depend on several factors; but openness to trade and direct investment and a skilful use of markets are definitely and important contributory factor.

As for inequality among nations, it is precisely those countries that embraced integration into the world economy, i.e. the Far Eastern Four and then the ASEAN countries, which raced ahead with dramatic growth rates whereas several countries of Africa, Latin America and Asia that looked inwards failed to deliver growth and also made little dent on poverty.

The evidence on trade and investment impoverishing our workers is also flawed. My own research suggests that the downward pressure on workers' wages due to technical change has been dampened, not magnified, by trade with the poor countries. Research also shows that big corporations use abroad technologies similar to those at home, instead of exploiting lower standards or forcing them yet lower through their financial clout.

One result of these mistaken arguments against globalisation has been an insistent clamour for certain environmental and labour standard to be linked to rules on international trade. But by seeking to create new obstacles to free trade, you undermine the freeing of trade, while mixing up trade with a moral agenda undermines that very moral agenda. It gives other countries the definite impression that you are using ethical rhetoric to mask protectionist self-interest.

The notion that global free trade and investment are responsible for poverty, inequality, lowering of standards and harming social progress is little short of astonishing. Yet national politicians and international bureaucrats give it who think that going along is way of getting along. In denying the virtues of globalisation, they actually harm the very causes they profess to embrace.

2

The Nation-State and Globalisation

The world has changed dramatically. Some of the changes are as yet only dimly understood. We are all going to be confronted with many challenges to the whole concept of government and to the role of the nation-state as we move into the next century.

There are two principal aspects to these changes. Globalisation of the world economies is sharply limiting the independence of action of the nation-state. In addition, we are only just beginning to understand what the existence of one superpower, supreme militarily, financially, means to the evolution of world diplomacy and world polities.

These remarks are directed to the first aspect. Governments are now losing influence. Private enterprise, capitalism, summarized as 'the market', is gaining power. Privatisation is a keyword. Across the political spectrum, liberal conservative and formerly socialist parities have all accepted the downsizing of government, the privatisation of many activities and the reduction of government debt. Governments in crisis in the developed or in the developing world have been left in no doubt about what they should do.

The International Monetary Fund and the World Bank have made it clear that assistance would not be available to countries in distress unless appropriate policies were put in place, and IMF prescriptions often involve substantial and detailed microeconomic reform within a country with considerable hardship for its people.

Meanwhile, competition for international capital has become much more severe. In the early independence years, Common Wealth countries believed they could write their own internal rules about the performance and behaviour of capital. Now those rules have to be rewritten to maximize international attraction. The relationship has to be competitive; the rules have to be friendly to capital. This is a totally different environment from the one in which most Common Wealth countries gained their independence in the immediate post-war years.

The new global organisation of industry has significant consequences for social policy. Many governments would have conducted policies designed to see that workers gained a fair share of the returns of an enterprise. With the globalisation of industry, such policies are no longer possible. Governments now tend to argue for lower wages, for smaller workforces, to maximize the competitiveness of their particular country as a home for global corporations. This has consequences of enhancing the profit share as opposed to the wage share of a particular enterprise.

One direct consequence of these changes is a significantly growing disparity in wealth between rich and poor in all countries worldwide. This may not matter so much if the poor were also becoming better off compared to their own earlier standards but in many cases this is not so. The idea of a living wage is no longer relevant. Workers in some countries are often paid a wage which could not support even the smallest of families. In this day, if that is what the market determines, then that is what must happen.

In today's world, governments must fashion their policies to meet the wishes of the international market place. There are fundamental differences from earlier times. The global organisation of industry in which national boundaries become irrelevant is certainly new. Some aspects of information technology can operate much faster and with more devastating effect than the old cable system of the last hundred years. This has led to an explosive growth in financial markets. The volume of money traded each day is huge (and) through modern communications, this finance has great mobility.

We all know enough of markets to know that they favour the powerful, the united and the strong and that markets can overwhelm and destroy smaller player. Sometimes smaller players are entire nations.

Those who suggest that the markets alone must be allowed to determine economic outcomes favour a world in which the large will do much better than the small. So far as countries are concerned, most Common Wealth countries are in the smaller category in a world in which large financial institutions and manufacturing corporations operating globally will dominate traded and commerce.

For most countries, banks and financial institutions, which are part of the culture of that country, will become a matter of the past. Banking services will be American, European, Japanese or perhaps Chinese. The consequences of this market dominances are clear. Corporations need a global spread and many national rules for the good order and conduct of business and commerce will no longer be relevant.

For the world as a whole, the most serious problem is volatility, possibly leading to systematic breakdown. Since the Asian economic problems of 1997, there has been a great deal of discussion about the present system and about changes that need to be made.

For a while it appeared that the United States really was going to mover the reform process forward but now the tendency seem to be 'its all right, we have escaped, leave well enough alone' ...There is a need to reform the system, to establish much tougher international rules for prudential supervision and control. The IMF has demonstrated time and time again that it is not interested in avoiding crises, it is only interested in picking up the pieces after they have occurred. If this is its charter, it certainly needs reviewing. The IMF's presents operations are inadequate.

Since governments have seemingly lost significant power to corporations and to financial markets and since they do operate within an increasingly globalised framework, individual governments are not capable of undertaking this task. The task is international and global. Whether it is a reformed IMF or a new institution is a matter for debate.

At their last meeting the Common Wealth Finance Ministers pointed to a number of changes, most of which are desirable, but there was no sense of great urgency, no sense of dynamism. They spoke of a need for new financial market architecture but nobody has tried to spell out what means.

There are two specific tasks: how to preserve some form of equity and reasonable competition in a globalised market place and how to establish stability within the financial markets themselves.

The IMF's financial resources should be strengthened as a means of averting crises through the provision of contingency funds. Immediate access to adequate funding can be essential for this purpose if crises are to be avoided. Finding a way to encourage the IMF to help avert crises instead of just reacting to crises after they have occurred is a most important requirement.

In any liquidity arrangement, assisting a country is distress, the IMF should take care not to absolve lenders of their responsibility. In some cases IMF should take care not to absolve lenders of their responsibility. In some cases IMF bailouts have done more to help the lenders than the countries themselves. The lenders need to carry their own risk.

For poor countries, how to protect themselves and advance the welfare of their own people in an unpredictable world is a major challenge and very often a major problem. Apart from moves to establish greater stability designed to avoid systematic breakdown within the world's financial system, there also need to be urgent moves to establish an international body to establish rules for fair trading in a globalised environment. Middle ranking and small countries would have most to gain from such an innovation.

3

Urbanisation and Globalisation

How we handle globalisation will determine whether our cities and our civilisation will be divided and violent or user-friendly and peaceful. We cannot get a clear picture of urban life in the 21st century, especially in the poor countries of the South, unless we take into account the phenomenon of globalisation, which has already brought dramatic changes make their first appearance. So it is there too that the great upheavals of the next century will take place.

Globalisation given shape to the "Global Village". He "information era" that it ushers in compress time and we are now living in a world speeded up as never before. Worldwide urbanisation is proceeding at a similar rate and its pace in the poor countries of the South seems terrifying. By 2025, two-thirds of humanity will be living in cities and towns, where the best opportunities in life tend to be.

Globalisation also accentuates a "new urban geography" in both North and South. Islands of rich consumers are springing up in cities amid an ocean of deprived people. More and more unemployed people, immigrants, minorities and the homeless, are pushed into cities by pressure from "market economies". As a result, all urban area—not just those in the poor countries of the South—will have to deal with growing internal tensions. In New York, for example, the poorest 20 per cent of the population earns 15 times less than the richest 20 per cent.

Cities have always had their smart neighborhoods and their dangerous areas. But such social and geographical segregation has changed in pace and scale because of the growth in the urban population, the increase in "illegal" migrant and rising uncertainty.

In fact, we have entered a period of historical transition, where discontinuities prevail over adjustment. Radical changes in the nature of production and jobs and the incredible concentration of capital in the hands of the financial sector and speculators weigh much heavier in our lives these days than the state's efforts to adjust and improve the market economy. Segregation in cities has been given a new lease of life whose consequences we do not know. It has reached unprecedented dimensions because of the explosive growth of urban areas.

According to one scenario, things will go badly. The growing pace of globalisation will increase uncertainty about the future. Fear and defence mechanisms will grow among people and institutions, fuelling intolerance, xenophobia and mistrust of everything new or foreign. Urban tensions will manifest themselves with increasing violence, and segregation will sharpen. Public areas will be abandoned and become dangerous no-man's lands, the wretched abode of society's rejects. Cities will lose their original function of being a crossroads for meeting and exchange.

If globalisation also continues to go hand in hand with deregulation of financial markets and an unchanged level of indebtedness of poor countries, the latter will not be able to maintain their urban infrastructures. And if on top of this there is corruption and lack of political will, challenges to the system will increase and violence will grow. Cash-strapped authorities will respond with undemocratic mafias which provide them with funds.

According to a second scenario, everything will be all right. In line with the principle that "everything the state does is public, but the state doesn't control everything that is public," a new social contract will be drawn up between the state, the market, the working population and civil society, including NGOs. Cities will develop a new quality of life by providing citizens with forum for exchange. Jobs will be created in the social sector, in the fields of the environment, education, research, culture and leisure, opening up possibilities for young people.

In the countries of the South, long-term development strategies will be drafted and urban planning practiced, taking advantage of the opportunities provided by globalisation but without falling into its traps. Town planning will become part of the political process, and the state will work with the private sector, monitored by institution of civil society. Adequate housing will be built with the help of micro-credit and controls on the price of building material. Improved infrastructures will enable marginal areas to become part of the civilized part of the city Democracy will come up with new ways of governing with the help of networks of involved citizens.

In a transitional scenario, action strategies should fall somewhere between these two extremes. They should include social goals so that in big urban areas a society emerges which is founded on participatory democracy and on "capitalism with a human face" or "market socialism".

But the outlook is less clear than ever. Let us hope the present transition will lead rapidly to a new revival of humanism, whose first signs we are already seeing. This would open up the road to a development which is fair, humane and peaceful.

4

Globalisation and Knowledge Divide

Globalisation looks very different when it is seen, not from the capitals of the West, but from the cities and villages of the South, where most of humanity lives. Four examples taken from India, illustrate how the paradoxical forces shaping globalisation look when seen from the other side.

Five school children died in a remote village in India after drinking water and powdered milk mixed in a vat that had contained a powerful insecticide. Nobody could read the label of the vat and the children were poisoned. The insecticide in question has been banned in practically every industrialized nation; its sale continues only in places like my country.

Secondly, an important annual event recently took place in North India. Potato growers gather there to exchange the best seeds they have produced in the last year. It is an act of pride for communities to share with others seeds that will help improve the production of potatoes. A transnational corporation attended the festival and are now working to patent the genes of these traditional foodstuffs in order to sell them as profit.

India's macro-economic indicators are excellent. In the offices of investment bankers, you will be told that India is a great investment opportunity. The situation is not so rosy, however. Thirty per cent of the population have been living below the poverty line for the last so many years. Ten per cent of the population are living below the critical poverty line: their income is insufficient to pay for event minimal nourishment. So much of the workforce is unemployed or under-employed.

A distinguished North American political scientist, Dr. Benjamin Barber, recently pointed out that in the United States democracy had degenerated into bringing one group of rascals in for four years, and then throwing them out and replacing them with another group of rascals for four years. From the perspective of the South, that looks very good! In a context where rascals manipulate elections and stay in power for fifteen or sixteen years, I would appreciate the chance to throw them out through peaceful elections every four years.

Thus, the complaints of the North are often the aspirations of the South. Progress in industrialized nations can be a threat to developing countries.

Ten year ago, in the euphoria of globalisation and the expansion of services and finance that followed the fall of the Berlin Wall, I advanced the idea that we were entering a fractured global order. Globalisation brings us into contact with one another, but it also strengthens profound divisions and fractures in terms of societies and income, and most importantly in our capacity to generate and utilize knowledge. Over the last ten years, the concentration of wealth and power has greatly increased both within and between societies.

There is a real risk of two civilisations emerging, with two ways of viewing and relating to the world: one based on the capacity to generate and utilize knowledge, the other passively receiving knowledge from abroad and deprived of the ability to modify it.

The world now faces the prospect of this Knowledge Divide becoming an unbridgeable abyss. We need the international community to return to the basic principles of international co-operation and introduce the idea that a minimum level of science and technological capability, including access to the Internet, is an absolute necessity for developing countries and should be the subject of international solidarity.

This can be achieved. However, contrary to the situation of 20 years ago, national governments are no longer the major players in the game of science and technology. Whether we like it or not, the private sector and the international community of scholar must be invited to the table with governments from the North and South to begin discussing an agenda for the mobilisation of a science and technology for development. United Nations with a mandate for the development of the sciences, has a special role to play in the revitalisation of international co-operation in this field.

5

The Truth About Global Competition:

The Economic Myths Behind Globalisation

Local communities everywhere are on the front lines of what might well be characterized as World War III. It is not the nuclear confrontation between East and West between the Soviet Union and the United States—that we once feared. It is a very different kind of conflict. There is no clash of competing military forces and the struggle is not defined by national borders. But it does involve an often-violent struggle for control of physical resources and territory that is destroying lives and communities at every hand. It is a struggle between the forces and institutions of economic globalisation and the communities that are trying to reclaim control of their economic lives. It is a conflict between competing goals—economic growths to maximize profits for absentee owners versus creating healthy communities that are good places for people to live. It is a competition for the control of markets and resources between global corporations and financial markets on the one hand and locally owned businesses serving local markets on the other.

Two things of fundamental importance to each and every one of us are now very much at stake.

- Will people and communities control their local resources and economies and be able to set their own goals and priorities based on their own values and aspiration? Or will these decisions be left to global financial markets and corporations that are blind to all values save one—instant financial returns?

- Will the life sustaining resources produced by the regenerative capacities of our planet's ecosystems be equitably shared to provide for the material needs of all of us who inhabit this bountiful planet, as well as for our children and their children unto the seventh generation and beyond? Or will we allow a global economic system that is now functioning on auto-pilot beyond conscious human control to consume and destroy the ecosystem and our social fabric in its insatiable quest for money?

Economists, politicians, corporate spokespersons and the media have for years been touting the benefits of the global economy. They have called on us to support trade agreements such as the North American Free Trade Agreement (NAFTA) and the World Trade Organisation (WTO) to remove the constraints of economic borders and open to everyone the opportunities of growth and prosperity in the global economy. They have promised rich rewards for those workers and communities that become successful global competitors.

Many of the most ardent boosters of economic globalisation met earlier in the year at the annual meeting of the World Economic Forum. This Forum for years brought together top industrialists and political figures from around the world to advance the proposition that removing tariffs and other restrictions on the free international flow of trade and money is a key to creating new economic opportunity and prosperity. It thus caused quite a stir when the Forum publicly announced that economic globalisation is producing disastrous consequences that threaten the political stability of the Western democracies. Their warning bears close examination for being one of the most honest and accurate assessments of the consequences of economic globalisation yet produced by leading advocates of that process. The observation is that:

- Economic globalisation is causing severe economic dislocation and social instability.
- The technological changes of the past few years have eliminated more jobs than they have created.

- The global competition "that is part and parcel of globalisation leads to winner-take-all situations: those who come out on top win big, and the losers lose even bigger."
- Higher profits no longer mean more job security and better wages. "Globalisation tends to delink the fate of the corporation from the fate of its employees."
- Unless serious corrective action is taken soon, the backlash could destabilize the Western democracies.

We don't have to go far to find examples of what they are talking about and why people are getting a bit upset as they wake up to the realities of who is winning in the ruthless competition of the global economy. The disparities between the winners and losers in the global competition are becoming more obscene with each passing day.

We are coming to realize that the extravagant promises of the advocates of the global economy are based on a number of myths that have become so deeply embedded in Western industrial culture that we have grown to accept them without examination.

- The myth that growth in GNP is a valid measure of human well being and progress.
- The myth that free unregulated markets efficiently allocate a society's resources.
- The myth that growth in trade benefits ordinary people
- The myth that global corporations are benevolent institutions that if freed from governmental interference will provide a clean environment for all and good jobs the poor.
- The myth that absentee investors create local prosperity.

The Growth Myth

Our measures of growth are deeply flawed in that they are purely measures of activity in the monetized economy.

Expanded use of cigarettes and alcohol increases economic output both as a direct consequence of their consumption and because of the related increase in health care needs. The need to clean up oil spills generates economic activity. Gun sales to minors generate economic activity. A divorce generates both lawyers fees and the need to buy or rent and outfit a new home increasing real estate brokerage fees and retail sales. It is now well documented that in number of other countries the quality of living of ordinary people has been declining as aggregate economic output increases.

The growth myth has another serious flaw. Since 1950, the world's economic output has increased 5 to 7 times. That growth has already increased the human burden on the planet's regenerative systems—its soils, air, water, fisheries, and forestry systems—beyond what the planet can sustain. Continuing to press for economic growth beyond the planet's sustainable limits does two things. It accelerates the rate of breakdown of the earth's regenerative systems—as we see so dramatically demonstrated in the case of many ocean fisheries, and it intensifies the competition between rich and poor for the resource base that remains.

This is vividly illustrated by many of the development projects in India many funded with loans from the World Bank and other multilateral development banks—that displace the poor so that the lands and waters on which they depend for their livelihood can be converted to uses that generate higher economic returns—meaning converted to use by people who can pay more than those who are displaced.

The Myth of Free Unregulated Markets

It is almost inherent in the nature of markets that their efficient function depends on the presence of a strong government to set a framework of rules for their operation. We know that free markets create monopolies, which government must break up to maintain the conditions of competition on which market function depends.

We also know that markets only allocate efficiently when prices reflect the full and true costs of production. Yet in the absence of governmental regulation, market incentives persistently push firms to cut corners on safety, pay workers less than a living wages, and dump untreated toxic discharges into a convenient river. In our present competitive context if management does not take such measures, they are likely to be replaced by the owners or bought out by someone with less scruples who will.

The Myth of Free Trade

Many so-called trade agreements, such as the North American Free Trade Agreement (NAFTA) and the World Trade Organisation (WTO) are not really trade agreements at all. They are economic integration agreements intended to guarantees the rights of global corporations to move both goods and investments wherever they wish—free from public interference and accountability. WTO is best described as a bill of rights for global corporations.

The Myth that Economic Globalisation is Inevitable

Many of the people who claim globalisation is a consequence of inevitable historical forces are paid to promote that message by the same global corporations that have invested millions of dollars in advancing the globalisation policy agenda.

The Myth that Corporations are Benevolent Institutions

The corporation is an institutional invention specifically and internationally created to concentrate control over economic resource while shielding those who hold the resulting power from liability for the consequences of its use. The more national economies become integrated into a seamless global economy, the further corporate power extends beyond the reach of any state and the less accountable it becomes to any human interest or institution other than a global financial system that is now best described as a gigantic legal gambling casino.

All over the world people are indeed waking up to the truth about economic globalisation and are taking steps to reclaim and rebuild their local economies. Such communities face basic choices as to how they will divide their efforts between competing for a share of the declining pool of good jobs that global corporations offer and working to create locally owned enterprises that sustainably harvest and process local resources to produce the jobs and the goods and services that local people need to live healthy, happy, and fulfilling lives in balance with the environment.

Our experience with the real consequence of economic globalisation is pointing to many important lesions. One such lesson is that economies should be local, rooting power in the people and communities who realize their well-being depends on the health and vitality of their local ecosystem. If it is protectionist to favour local firms and workers who pay local taxes, live by local rules, respect and nurture the local ecosystems, compete fairly in local markets, and contribute to community life—then let us all proudly proclaim ourselves to be protectionist.

Such choices are not isolationist, to the contrary, they create a foundation for creative cooperation with our neighbours, whether they be in the United States or in other countries, to share experience, ideas, and technology, and to join in international solidarity in rewriting the rules of the global economy to favour local over global businesses, and to encourage cooperative relations among people and communities. It is our consciousness, our ways of thinking and our sense of membership in a larger community—not our economies—that should be global.

Millions of people are also making an important discovery that life of is about living not consuming. A life of material sufficiency can be filled with social, cultural, intellectual, and spiritual abundance that place no burden on the planet.

It is time to assume responsibility for creating a new human future of just and sustainable communities freed from the myth that greed, competition, and mindless consumption are paths to individual and collective fulfillment. It will take millions of people around the world linked together into a powerful political coalition aimed at radical, political and economic reform to win the war that global capital is waging against us.

6

High World Trade Growth vs Output—WTO Sees Link to Globalisation

World trade in merchandise goods is expected to increase in volume by 8 per cent in 1995 down marginally on the very high $9^{1/2}$ per cent for 1994. Although the current outlook is for a further modest slowing next year, trade growth will remain above the average of the past decade.

Recent trade growth figures continue to exceed world production growth by a large margin in 1995 probably by a factor of almost three and next year close to double. This persistent pattern relates closely to the "globalisation" of the world economy; a process which, brings far-reaching benefit and which can be promoted through the further development of the multilateral trading system.

The recent growth is a follows:

- A 13 per cent rise pushed the value of world merchandise trade past the $4,000 billion mark for the first time, to $4,090 billion.
- An 8 per cent increase in the value of trade in commercial services to $1,100 billion after near stagnation in 1993;
- A 23 per cent increase in the dollar value of merchandise trade in the first six months of 1995 which, allowing for the depreciation of the US dollar, is consistent with a full-year growth in trade volume of 8 per cent.

Globalisation

Over the period from 1950 (when the process of trade liberalisation through the early GATT Round got under-way) to 1994, the volume of world merchandise trade increased at an annual rate of slightly more than 6 per cent and world output by close to 4 per cent. Thus, during those 45 years world merchandise trade multiplies 14 times and output $5^{1/2}$ times. However, the excess of trade growth over output growth varied; from an average of a mere half percentage point in the period 1974-84 to nearly 3½ percentage points in the most recent 10 year. In fact, the excess during the years since 1990 has been much higher still but it is not yet clear whether or not this represent a permanent shift to a faster rate of increase in the world's trade-to output ratio.

To the question "will globalisation continue?" In this regard one has to observe two factors—technological change and the evolving strategies of firms and individual investors—impart a natural momentum to global integration. It is government policies which can speed-up, slow down or even reverse progress on global integration. In this context, the role of non-discrimination—in particular, through the "most-favoured-nation" (MFN) clause—is examined.

The MFN Clause

MFN was the centerpiece of a multiplicity of bilateral trade agreement reached in Europe in the second half of the 19th century, a period marked by very low tariffs and rapidly increasing trade. In contrast, the 1920s and the 1930s saw efforts to restore liberal trade through international trade conferences rather than legally binding commercial treaties based on MFN. The failure of these efforts contributed to the Great Depression and provided some of the roots of military confrontation in 1939. It was only after the War that negotiations established what became the GATT, a multilateral contract consisting of rules and disciplines and based firmly (Article) on MFN treatment.

The GATT system has been a post-war bulwark against a return to the trade chaos of the 1930's. In the 1990s, a disintegration of the globalized international economy on the scale of the 1930s is almost unthinkable. In contrast, today "the threat that would be posed by a loss of credibility of the multilateral rules" (now represented by the WTO) would be "a fracturing of the global economy into inward-looking and potentially antagonistic trading blocs".

One can suggest two safeguards against such an eventually:

- The examination of new ways to ensure that free-trade areas and customs unions remain outward-looking and complement rather than compete with the multilateral trading system; and
- Progress in dealing, at the multilateral level, with new issues tied directly to the further evolution of the global economy. These include telecommunications, financial services, environment, competition and investment polices among others.

Progress in dealing with these and other issues at the multilateral level will have a significant impact on the future pace of global integration, both directly and through its impact on the credibility of the multilateral system in influencing the broad spectrum of national trade polices.

7

Renewing the State

Many view globalisation as a technology driven global order that has led to an intensification of interconnectedness among nations. This, however, is merely one fact of globalisation, and does not presuppose the ideological homogenisation or the rapid retrenchment of the welfare state that is currently underway.

The dispute over globalisation is not about the intensification of global interconnectedness. Rather, it is over the vision of the global system that globalisation projects. This vision entails a global economic system with identifiable rules of behaviour in trade, finance, taxation, investment policy, intellectual property rights, and currency convertibility, all of which are crafted along neo-liberal principles with minimal governmental regulation. This global system represents a new phase of capitalism which is "more universal more unchallenged, more pure and more unadulterated than even before".

For many critics, globalisation is essentially an anti-democratic process that excludes the interests of a wide range of groups. But the process is not shaped by market forces alone. It is only made possible by the acquiescence if not active support of governments, especially those in advanced countries.

Governments in developing countries, meanwhile, are often said to be unable to stand up to gloalisation without incurring severe costs. The government of South Africa, for example, could be punished by capital flight if it insists on implementing its agenda of social reform. The masses of South Africa, however, are likely to sustain heavier costs if the government abandons its reforming mandate. Faced with such a dilemma, governments have generally selected the side of capital for a simple reason.

The list of problems caused by globalisation is long. In low-income countries such as those in Sub-Saharan Africa, where governments have been unable or unwilling to provide their populations with even the most basic protection from the new phase of global capitalism and structural adjustment programmes, the people's plight has been particularly severe.

Opponents of globalisation are addressing genuine problems. But it is uncertain whether they will succeed in reversing gloabalisation or even in mitigating its adverse impacts. To begin with, many of them are badly organized. Most of them have also rallied around specific issues instead of articulating a comprehensive counter vision. At this point, the counter vision they project appears to be a global system which is not shaped by the narrow interests of capital but which accommodates the interests of diverse social groups. This vision, however, is not yet well developed.

Further more, these opponents have yet to develop viable strategies to constrain globalisation. Some argue for weakening or even abolishing institutions such as the World Bank, the International Monetary Fund, and the World Trade Organisation, which they view as agents of globalisation, it is unclear why business interests and governments would allow this to happen. The relevance of these bodies is only likely to decline if Third World countries, especially middle-income ones, begin to reduce their dependence on them under pressure from their populations.

Yet the main problem faced by these critics is that many of them do not see the relevance of the state. A successful struggle for genuine popular democracy can liberate the state from the grip of corporate and financial interests, turning it into a critical agent for the promotion of broad social interests. Many NGOs rely instead of civil society, though this cannot substitute the state in policymaking. The struggle against globalisation is essentially a struggle for democracy; the state cannot be bypassed, but must be won.

Free Trade as Peacemaker:

The Benefits of an Open World Trading System

Globalisation by free trade according to the principles of the World Trade Organisation (WTO) offers the only realistic opportunity to integrate the world peacefully and in time to prevent a major disaster. The primacy of the economy over politics is the most important vehicle for a successful world domestic policy.

Since Adam Smith, traditional economic theory has on principle been well-disposed towards free trade. Free trade enables better use of the world's economic resources than does national protectionism. Countries can concentrate on their respective strength and draw from their trade partners the goods they need, but do not produce. But there have always been objections against free trade.

The international trade system has always been encumbered by disparate accusations of unfair competition. The fear that foreign competitors use unfair methods, such as dumping, as and is widespread. If one were to believe all the charges of dumping that are made, then international trade would have been completely destroyed long ago. Great restraint should be exercised with respect to allegations of dumping if one is interested in maintaining an interweaving of international economic activities.

The Free Trade Oppsition Cloaks itself in Dumping Charges

The modern form of the struggle against free trade cloaks itself in the accusation of ecological dumping or social dumping. With this difficult subject matter, one should not make sweeping generalisations. These things also are not gone into in detail in what follows.

Environmental protection is an asset that every economy produces at the cost of other assets. The people's preferences for the asset of environmental protection probably varies from country to country. It is also completely legitimate and does not at all distort trade if the environmental provisions-in line with the different national preferences—vary from country to country.

In the rich western European economic region, one should guard against a new form of cultural imperialism. It is not for this part of the world to impose its preferences for environmental assets on other countries, especially Third World countries. Free world trade brings not only economic advantages. Even more important is its contribution to lasting world peace.

In view of world population growth, every standstill in the movement towards a peaceful world society must be seen as a step backwards. We are compelled to run a race between the growing problems and the development of stable institutions to overcome them peacefully at global level. Economic history since the end of the Second World War shows clearly that free trade under the old GATT was of decisive importance for the prosperity of the industrialised nations.

The principle of help for self-help has nowhere been applied so consistently as on the free world market. In reverse, the examples of countries that cut themselves off from the world market show the disastrous consequences of the rigidity of a society which shuns the pressure of international competition.

Revolutionary Success of Open-Market Policies

The West's policy of open markets pursued since 1948 and reinforced since 1989 has led to a dynamism which, in the true meaning of the word, is revolutionary. More than half the word population now lives in countries with annual

GDP growth rates of more than 5 per cent. Europe is not among that group, which may be why it also stands somewhat apart in its mentality.

Certainly, there also can be undesirable trends in free trade. There is not ideal systems; one must choose between imperfect potentialities. However, no realistically better substitute for the free trade system is in sight, not even with respect to the goals of a pacified world: an ecological sound world economy and a balance of global dimensions between the poor and the rich. An ideal government of philosopher kings armed with absolute power certainly could do some thing better than does free trade—but such a government remains fictitious. There are tangible and narrow limits to what the political system, whether democratic or not, can effect in a positive sense. This is how the structural conservatism of democratic and other political systems impedes the timely assertion of reforms necessary to achieve a world peace society.

The GATT was turned into the World Trade Organisation (WTO) a few years ago. Besides extending the free trade principle to services and additional agricultural sectors the new agreement foresees above all the full inclusion of the Third World in the system. The agreement commits the industrialised nations to open their markets to developing and threshold countries.

Other important points are the strengthening and tightening of the dispute mediation process. Based on a system of relatively independent ad hoc panels, it permits complaints against WTO member countries for violations of the agreement. Thus, what is arising here is an effective global jurisdiction within the meaning of a peaceful world domestic policy.

Exclusion as Penalty

The decisive sanction mechanism of the WTO—which is not a specialist organisation of the United Nations—is the threat of exclusion. Exclusion would deny the penalised country free access to the markets of WTO members on the basis of most favoured nation status. This is a threat that requires no armed force, but is very effective. No country can still afford to do without the beneficial effects on prosperty that participation in international trade brings.

Thus, with the threat of denial of access to world markets for violating WTO rules, and the guarantee of a more or less fair competition for a country's own products for abiding by them, a non-military sanctions system has come into being. That is substantial progress on the path to a pacified world.

Certainly, this sanction system's sphere of influence is limited for the time being. Essentially, it will be used to assert the game rules of free trade. It offers no legal grounds for pressing other goals, such as on human rights. Attempting to expand it in this direction would for the foreseeable future put the entire system at risk.

In the current debate on globalisation, the question arises of whether the world economic institutions should not be converted in this manner, that politics regains its autonomy, and that the primacy of politics can be restored. The critics of globalisation point to the constraints to adjust which the world economy exercises on national or continental politics. However well this demand for the primacy of politics may be justified in philosophical terms, it virtually comes down to a demand for the ascendancy of the conservative principle.

Danger of a Slowed-Down World Integration

The demand for the primacy of politics is gaining strength from the desire to avoid the pressure to adjust which the dynamics of world events are exerting. It is today a conservative, and in fact a reactionary, longing for the (Utopian) return of the functioning European welfare state of two or three decades ago. If it were asserted, it would mean practically slowing down world integration. It would run dead against the goal of a world policy based on a desire for peace.

The present primacy of the economy over politics—in terms of the free movement of goods, services and capital—

is basically nothing more than the priority of the global principle over the provincial, the national principle. As such, it gives the principle of change pre-eminence over the principle of maintaining the status quo. What gives the primacy of the economy its legitimacy? Probably not the thought that world peace and better be secured by this means. Its legitimation lies in the very indirect economic success that the free trade system delivers. For the reflective observer, the question remains of whether this legitimation is sufficient.

To answer this question, however, and particularly if one pleads for maintaining the ascendancy of the economy, it appears appropriate to outline the consequences that can be expected from further integration of the world economy. As can be seen today in East and Southeast Asia, the growth dynamics of the world economy will lead to a marked rise in the living standards of a large part of the Third World.

Do Not Exclude Poor Countries from the Competition

The global consequences of Asia's growth should not have been seen only negatively. While it also may mean, for example, a great burden on the global climate, it leads at the same time to an acceleration of the process of falling birth rates and thus to an earlier stabilisation of the world population. Prosperity for the Third World is so far the only realistic answer to the urgent problem of population growth. And free competition on the world market in the only reliable means of achieving this prosperity in the course of some decades.

Despite ecological sacrifice in the medium term, continuation of Third World growth is the only way to solve the long-term ecological problems. One also should not forget that only those who can eat their fill and have a roof over their heads are prepared to reflect on ecology and discuss it.

As for the rest, the balance between rich and poor is more acceptable when the poor become richer than when the rich become poorer. That applies also at the international level. The market and access to it are peaceful sanctions of the world economic system on the basis of free trade. Those who are hungry and have nothing more to lose are more of a danger to world peace than those who have eaten their fill. The ruse of covering up domestic problems by cross border military aggression will become less attractive to the degree that a country's own economy is integrated in the global economic system. The more countries are economically dependent on each other, the more unlikely it is that they will wage war on each other.

Globalisation by free trade according to the principles of the WTO offers the only realistic opportunity to integrate the world peacefully and in time to prevent a major disaster. The primacy of the economy is the most important vehicle for a successful world domestic policy.

9

Democracy and the Market Economy

Today the idea of democracy is triumphant; the model is in principle embraced in most countries the world over. You may say that the very word democracy has been hailed and misused earlier in history. The most repressing and totalitarian regimes have tried to mask themselves as "real" or "peoples" democracies. What has happened, however, is a historical demasking of these false pretences.

What exactly do we mean by democracy? There is now a general agreement that democracy cannot be defined by purpose or policy or levels of mass mobilisation. It must be defined as a political system where different parties or individuals compete for power through regular free elections where all adult citizens have a vote. Moreover, a democracy must uphold certain basic human rights and well-defined freedoms which make the political process possible, and respect the opinion and integrity of the individual. No other definitions hold, and we should be careful when we talk about "real" democracy versus "formal" democracy. A society, which in real life upholds the constitutional or formal democratic principles and which in practice applies the rights these principles imply, is by definition a democracy. A society with a beautiful-sounding constitution but where none or few of these rights are respected is certainly not a democracy.

Democratic Government no Guarantee for Equality

It is important to understand that democratic government does not necessarily mean good government in the sense that those in power make wise or well-considered decisions. Nor does it mean that conflicts inherent in the society are reduced to a minimum. Demands for democracy, social justice and a better life have historically gone hand in hand, but this does not mean that the establishment of a democratic system actually does lead to an improvement in social conditions or equality. It is also quite clear that some societies have a sort of outer shell of democracy but in reality, exclude large groups of people from having any political influence whatsoever. The actual differences in living conditions are so enormous and so entrenched that these people have no confidence at all in the political system even if it is democratic according to the definition. In these cases—for example in some Latin American countries—one can talk of a "masked hegemony with competing elites" where the outcome of struggles for power has little relevance for the masses. It is a sort of social and political half-authoritarian system—but disguised as a democracy—where the military often have a significant influence.

In the rhetoric of the day the term market economy and democracy are used as if they were synonymous or at least naturally emerging at the same time. But this is wrong—or at least misleading. When the market economy or capitalism finally established itself in the 1800s and came to characterize modern industrial civilisation, democracy was at best in its infancy. In fact one could argue that democracy grew out of the contradictions and social dynamism inherent in the market economy of the capitalistic system. In this century we have a long list of terrifying and repressive regimes, which have nevertheless upheld the virtues of a market economy. That some of these regimes have for ideological and security reason been hailed as bastions against communism, and also dignified members of the so-called free world does not transform them into democracies. In this company it is perhaps unnecessary to remind ourselves that the colonial system was assuredly not democratic, but was certainly based on capitalistic or market economic principles. It is the sad but irrefutable historical coupling between Western democracy, colonialism, and the plundering of resources in the name of the market economy which for understandable reasons meant that many of the leaders of national liberation movements looked for other models for the development of their young nations. In this connection it can be worth remembering what Nelson Mandela said soon after his release from 26 years of prison in the market economic but racist state of South Africa. "When we in ANC during 40 years struggled for democracy we were put in prison by the same people who are now telling us how we should behave to promote the democracy we have been rejected by all these year".

While we can see that a market economy does not automatically lead to democracy, a functioning democracy—as we have defined it—does seem to require some form of free economic system.

Democracy and Economic Freedom

Theoretically, it is conceivable that a political democracy could be combined with an economy totally controlled by the government—but experience has shown this to be very difficult. One could even argue that it is by definition impossible since democracy implies a certain freedom of economic choice and independent economic actors. A functioning democratic system presupposes what is now often referred to as a civil society—in practice, independent institutions, companies, organisations, the media etc., regulated by law but not subject to or controlled by those in power.

We must also see clearly that there are no unambiguous relations between economic growth, development and democracy. Democratic governments are neither very successful when it comes to structural reforms which may be to the disadvantage of important interest in the society, nor when it comes to welfare. The developing countries which have achieved the greatest success economically and socially over the last 20 years are the East Asian countries—which all have had various kinds of more or less authoritarian systems.

However, that does not mean that you can use these countries as models for the rest of the world. There is no globally valid link between an authoritarian form of regime and economic development, not even when development is defined only in terms of autocentric growth. Many social scientist—have tried to find some systematic connection between what we call development or modernisation on the one hand, and the political system on the other—but all have failed.

It is also obvious that one of several pre-requisites for economic growth and development is legitimate and reasonably well functioning government and governance. If the free market is to be a motor for development and improved welfare, and not just a meeting place for robber barons, the mafia and speculators, you must have a regulating and supportive state. If economic history teaches us anything, it is just this. Consider the astounding development in Germany after the war, or in Japan and the other East Asian countries some years later. There are many differences, but what they have in common is a well-functioning government apparatus with a long tradition.

Today we find ourselves in a historical situation where a large number of countries in the former communist states of Europe, in Africa, Asia and Latin America are at one and the same time trying to establish a new democratic system and new economic mechanisms. The situation is unique, and the intrinsic problems are unprecedented. Democracy as an idea has triumphed but in its practice it is in profound trouble. It is no exaggeration to talk of the crisis of democracy.

The former communist countries are certainly in crisis. As by-product of the past regimes, there is an intensive suspicion of the political institutions, of the state and the parties-and in this way also the legitimacy of democracy and the ability of the politicians to deal with the fundamental problems of society has been undermined. The lack of a democratic tradition is not overcome from one day to the next.

Many of the developing countries have similar difficulties. The introduction of a multiparty system does not in itself mean that one can manage the conflicts and social problems in a democratic way.

Countries in Transition

Both in the East and the South countries are trying, at one and the same time, to change the political and economic system. When the whole society is convulsed by economic changes, and where peoples' living conditions fundamentally change, it is not easy to develop and maintain a political system based on compromise and respect, including respect for minorities.

As in previous history the deep crises of legitimacy and general frustration feed national and ethnical conflicts. These conflicts establish themselves in societies where the authoritarian system, economic crises and the break down of traditional values rob people of any kind of kinship other than ethnical.

We cannot avoid seeing disturbing signs of this crisis of democracy also in the so-called "established democracies" of the rich countries.

It is obvious that the state of democracy varies from country to country as do the reasons for a feeling of dejection. But there are some similarities too.

The continuing and noticeable internationalisation limits the national freedom of political choice, available alternatives, and makes it more difficult for people to see the connection between "politics" and their actual living conditions. The governments are restrained by international economic events. The reaction of the stock exchange may be more important than that of the voters. The election results influence the stock exchange prices—but is it perhaps not also so that the stock exchanges, indirectly, also influence the election results? People feel themselves to be the victims of major economic changes, but no one seems to be responsible and they themselves feel they have little chance of influencing the outcome. The absence of clearly identifiable alternatives between the larger political parties provides opportunities for the populists and the extremists.

There is indeed reason to reflect on the lessons of the history of our turbulent and cruel century.

Priority for Growth

There is today much concern about the lack of resources for such urgent needs as the reconstruction of the East, a concerted attack on poverty and human development in the poorest countries, and environmental investments of all kinds. If the growth of world output returns to the levels of the 1980s, total output would grow by about one trillion dollar a year. There is, infact, no other way to resolve the economic and political crises multiplying in the world community than to give priority to the restoration of growth.

We are certainly not at the end of history as someone has argued. We are rather at a dramatic turning point, a moment of many possibilities and many dangers. What we do now, for a few years ahead, may direct the future for several decades—like the dramatic and fateful years immediately after the second World War. All nations, all governments, have a responsibility. The rich world has a special responsibility, not just moral because of its enormous economic and political power.

10

Myths and Illusions

The ride of precarity is rising, so that people who have never been poor no longer regard poverty as a distant prospect but as one so close that it could engulf them at any moment.

In 1989, the fall of the Berlin Wall was rightly welcomed because it marked the collapse of a system that provided a degree of equality but rejected freedom. Today there is a strong possibility that the system gradually spreading all over the world—a kind of neo-liberal fundamentalism—may also collapse. In its obsession with freedom, vital though freedom is, this fundamentalism disregards equality, a term which should not be regarded here in purely static and statistical terms, but as something dynamic and ethical. Equality can only be truly practiced in a context of social solidarity or to borrow from the vocabulary of the French Revolution of fratemity.

On the one hand, we have a world that is immensely rich in resources, possibilities, knowledge and experience; its constituent societies are freer and more dynamic than ever. There is an extraordinary potential for everyone to live a better life. But at the same time, new and ever higher walls are being built both between peoples and between social groups within individual countries. We are experiencing a travesty of development, which is creating a world bipolarized into extremes of wealth and poverty.

The most common reactions to this disastrous situation are very often the result of two misapprehensions. The first can only be described as ideological or doctrinaire since it is not based on the facts as they can be observed. It says that since the dominant system of values and things is by definition more than satisfactory, the persistence of impoverishment is merely a temporary blip. Enough time has elapsed, however, for us to see that this is not the case, including in countries where this system has been part of the established order for more than a century. One statistic is particularly eloquent. In just over 30 years, world production has approximately doubled, but the gap has more than doubled between the income of the 30 per cent of world's people living in the richest countries and the income of the world's poorest 20 per cent, according to the United Nations Development Programmes.

The second misapprehension stems from another form of blindness and illusion, namely the belief that poverty can be regarded exclusively as a moral issue, as if it had no other kind of implications for those who are not poor. Globalization is, however, a two-way process. It enable the countries of the North to export their values and their paradigms as well as their goods and capital to the countries of the South, but it also makes them much more vulnerable to the backlash of crises that afflict these countries. Even in the North, the cult of competitiveness is undermining situations once considered extremely stable. The tide of precarity is rising steadily, so that people who have never been poor no longer regard poverty as a distant prospect but as one so close that it could engulf them at any moment.

Because of inadequate socio-economic development, the extraordinary upsurge of democracy over the past 30 years remains a very fragile process, and there is a risk that the trend may be reversed. When hunger, disease and ignorance prevail, citizens' participation in decision-making becomes either non-existent or a mere charade. Democratic institutions become empty shells, representational bodies existing in form only and devoid of real significance

Social divisions caused by economic distortions exacerbate the failures of democracy which in turn pose serious threats to civil order within countries and to peace between nations. It is high time to face these obvious facts.

Growing Complexity in International Economic Relations Demands Broadening and Deepening of the Multilateral Trading System

By pressing on with liberalisation, by successfully providing a way forward in areas of trade where protectionism had long proved intractable, and by boldly addressing entirely new but very important aspects of trade, the Uruguay Round made a signal contribution to international trade relations. It was a landmark achievement to create he World Trade Organisation (WTO). But following any birth, the offspring must be nurtured. There are three major challenges to this new institution in the years ahead. The first is to consolidate what it did. The second is to give substance to this built-in negotiating agenda, which essentially constitutes unfinished business emanating from the Round. The third is to meet the new challenges already gathering on the horizon.

First, consolidation, or implementation. The sheer range of subjects that were covered in the Uruguay Round is daunting for even the hardiest trade hands. The texts of the results comprise no less than 19 Agreements, 24 Decisions, eight Understandings, and three Declarations. Some of these texts are obviously more important than others, but together they represent nearly 500 pages of carefully crafted language, replete with commitments. For some countries, a number of these commitments will coincide with existing policies. In other instances, they call for change. A concerted effort is required by all WTO members to consolidate the Uruguay Round results, and ensure full compliance. It is an open question whether phase-in arrangements for some of these commitments should be speeded up. Why the benefits of liberalisation in any country should be delayed one day longer than absolutely necessary. Even as they are, the commitments require steady, continuing work in national capitals and in the WTO on a day-to-day basis. It is activity which seldom catches the headlines, but it is essential to the proper functioning of the system.

However, the biggest, short-term, priority is to make sure the new dispute settlement system works in a legally and politically credible manner. When difficulties and disagreements are, the WTO's consultation, conciliation and dispute settlement provisions can be called into action. A willingness to abide by the dispute settlement procedures and findings, is just as important as respecting the rules. First, governments are making use of it in a manner which demonstrates considerable faith in the WTO. Around 20 cases have come to the Dispute Settlement Body-a number far greater than in any single of the GATT's 47-year existence. Second, the rapid automatic procedure together with the knowledge that at its conclusion the system is enforceable seems to be concentrating minds and encouraging quick settlements though the initial consultative process—the recent Us-Japan dispute on cars and spare parts is one of these cases. And that is the objective—to resolves trade disputes quickly, not primarily, to generate jurisprudence. Of course, many disputes will run their full course, and no doubt that will be able to produce objective, clear, well-argued judgments which will command the confidence of governments and legislators everywhere. Nobody need have any fear of arbitrary conclusions or a lack of neutrality on the part of WTO dispute panels or the new Appellate Body.

For all countries, new and detailed obligations have been created to notify policies and measures, so that trading partners can be confident that they have full knowledge of each others' policies Transparency is an essential ingredient for fostering mutual trust and encouraging respect for the rules. Indeed, one of the results of the Uruguay Round was the creation of trade policy review mechanism, where the trade policies of individual WTO members are examined multilaterally by turn, and in depth. These examinations provide an opportunity for countries to hold frank and non-litigious exchanges of view about each others' policies. They are a valuable contribution to transparency, and help to raise awareness among trading partners of policy issues.

In previous multilateral trade negotiations, unfinished business tended to reflect failure to agree on quite fundamental issues, such as whether to do anything about agriculture, or textiles, or whether to redesign the rules on safeguard measures. This was hardly the case in the Uruguay Round. However, by the end of negotiations in 1993, it was clear that extra time would be needed in a few key sectors. This is clearest in the field of services, where the already held post-Uruguay Round negotiations on trade in financial services and the movement of natural persons, and are in the midst of negotiations on the opening up of basic telecommunications and maritime transport services.

The negotiations on basic telecommunication are to be completed soon. They will open up significant new trade and investment opportunities. The negotiations coincide with industry trends towards liberalisation, attributable both to pressure from user industries and rapid technological development. But there is nonetheless resistance to the eradication of monopoly supply arrangements in many countries, and concerted multilateral action offers the best hope of securing far-reaching results. Success in these negotiations will mean that telecom operators should be able to offer a broad spectrum of competitively priced services, in both national and international markets. The United Nations is the vanguard of this negotiation, with one of the most liberal and low-cost telecommunications markets in the world. This is why its commitment to a genuine multilateral result is of vital importance. We need a strong result from WTO negotiations if we are to make the vision of the Global Information Society a reality—with all that it will mean for revitalizing economies, transforming our societies, and empowering people.

The negotiations on maritime transport services, on the other hand, deal with one of the most ancient means of exchange among peoples, one which retains its fundamental importance for the flow of merchandise trade. The prodigious improvements in shipping technology over recent year need to be matched by improvements in the policy environment in which these ships sail. This also is a negotiation where there are some firmly held positions, and it is essential that we keep on recalling that it is every bit as valid and important as the negotiations in other areas.

Another part of the Uruguay Round's unfinished business is the built-in agenda for future work. This comprises several elements. WTO members have already established a mandate to enter into successive round of negotiations in trade in services, with a view to achieving progressively higher levels of liberalisation. The first such negotiation must begin within five years. Similarly, in agriculture members are committed to engage in negotiations aimed at further reductions in agricultural support and protection. The time frame envisaged is the same as that for services. These commitments and a numbers of others in the WTO Agreement clearly reflect recognition of the need for continual, incremental trade liberalisation—a virtuous circle of global cooperative efforts that is the basis of an effective multilateral system.

12

World Trade—The Next Challenge

On 15 December 1993 the world changed. My be not as dramatically as the moment when the Berlin Wall fell, but then unlike that very necessary demolition job, the success of the Uruguay Round was a work of construction. Like the destruction of the wall, though, its effects will be profound and lasting ones felt far beyond its immediate context. It will be seen as a defining moment in modern history.

The importance of the Round can be seen in terms of boost it gives to job creation; to development; to investment; to economic reform; to the rule of law and in many other ways besides. All of these benefits are real and important. But the true value of the whole is much, much more than the sum of these parts.

Put simply, governments came to the conclusion that the notion of a new world order was not merely attractive but absolutely vital; that the reality of the global market–whatever ambitions some of them may retain for regional Integration–required a level of multilateral cooperation never before attempted.

No Losers in the Round

It has created a revolutionary framework for economic, legal and political cooperation. But now turn to the immediate results of the Round. Seeing them as a profit and loss account or a scorecard of winners and losers is to see them in static terms, as one-off conclusions with finite effects. This misses the point completely.

Every nation now needs an effective trading system, but especially so that small and poor. They have it. Everyone will also gain from the huge package of market access results even if they did not get every concession they were seeking from trading partners—it is the biggest market access deal ever negotiated.

However, the essence of the Uruguay Round's achievements is that they are dynamic. The new agreements, the new rules and structures it sets up—all mean a commitment to a continuing process of cooperation and reform of which the agreement in December was only the beginning.

Maintaining the liberalizing momentum will call for continuing effort and vigilance by participating countries. But now their energy can be focused through the Round's greatest innovation; the new World Trade Organisation (WTO) in place of the improvised basis on which the GATT has operated for 45 years, trade will now have a permanent forum appropriate to its importance in the world economy.

Technically speaking, the WTO will oversee the implementation of the Round's results, administer all the agreements in goods, services and intellectual property, and manage the unified dispute settlement system. But beyond these administrative functions, it will raise the political profile of trade a profile which has already been lifted greatly by the Uruguay Round. The WTO will have regular instead of occasional—direct Ministerial involvement. It will have a clear mandate to act as a forum for further trade negotiations. Most of all it will complete the transition from a trading system which largely restricted itself to policies at the border to one which also covers most aspects of domestic policy-making affecting international competition in goods and services, as well as investment.

Through the WTO, the Round will change the way the world economy is shaped. But it is not the final victory over protectionism and unilateralism. Any premature rejoicing would have quickly been cut short by the evidence since 15 December that major economic powers are still ready to take the unilateral approach to trade problems. Arguments for protectionism based on the alleged threat of low-cost competition to production and jobs will not just fade away because the Round is a success. The seductive appeal of "beggar-thy-neighbour" policies is highlighted by the seemingly greater vigour of the lobbies for protectionism than the advocates of open markets.

These dangers—and the speed with which they have resurfaced—make the achievement of the Uruguay Round all the more important, and its successful implementation all the more urgent. Implementation requires more than mutual backslapping about what we have achieved. It requires now that the US, EU and Japan, in particular, rapidly obtain final authority to ratify and also take a lead in providing the WTO with the means to fulfill its mandate.

The success of the Round has come at a time when it is even more vitally needed than anyone could have guessed when it was launched in 1986. Old structures and alignments have been turned inside out in trade as in every other area of international relations. We face a world of change and challenge, in which the reinforced trading system will be a primary source of stability and security.

The developing countries including India have become enthusiastic supporters of the multilateral trading system and the Uruguay Round every if all their demands were not met by industrial countries. The reasons lie in the changing economic policies of many developing countries and the clearer appreciation of the value of the GATT system that has grown along with these changes.

The challenge of new issues in world trade will be a major one for the WTO. The new organisation has to consider issues such as the links between trade and the environment, international competition policy, trade and investment, and trade and labour standards. To say a few words about trade and the environment since it is one area in which GATT member countries have committed themselves already to a comprehensive new work programme. They decided on 15 December, in conjunction with the adoption of the results of the Uruguay Round negotiations, to draw up a work programme on trade and environment by the Ministerial meeting in Marrakesh. Environmental policy-making is one of the most rapidly evolving areas of national and international policy-making, and it is entirely appropriate that emphasis should be placed now in GATT/WTO on ensuring better policy coordination and multilateral cooperation over the linkages between trade and environment.

Permanent Negotiations

The Uruguay Round may well be the last of its kind, but this in no way means the end of multilateral trade negotiations. On the contrary, it means they become a permanent event. Ad hoc negotiating rounds were necessary mainly because the GATT lacked the mandate or the institutional basis to operate the multilateral system to the full on a continuous basis. Between rounds the GATT has tended to lose momentum, often at the very times when it was essential to make the most of the liberalising impulse. This has allowed protectionism and unilateralism to recover and regroup and meant that each round has to start by regaining lost ground.

The positive results of the Uruguay Round will redefine much more than assumptions about trade. If they are exploited with the same determination, courage and commitment that went into concluding the Round, they should mean nothing less than a new start for sustainable growth and a new system of collective economic security for the world.

But if the trading system is now up to the job of supporting multilateral cooperation on such a wide scale, do the other structures of economic cooperation still meet the bill? The establishment of the WTO will put trade and investment on a par—perhaps rather in advance—of cooperation in monetary and financial areas. The WTO will stand alongside its original Bretton Woods sisters, the IMF and the World Bank. The three institutions must learn to work together even more effectively and closely. For example, rather than each body conducting separate reviews of country policies, is there not a case to be made for a more integrated approach on country reviews? But that does not, on its own, add up to effective multilateral economic cooperation. The question really has to be asked seriously: are the G7, the OECD, the regional groupings adequate to provide that cooperation?

It is the next challenge of international economic leadership—the challenge of translating the common interest in global growth into a practical and effective mechanism for solving our common economic problems together. So, the Ministers meeting in Marrakesh is an historic event which will establish the World Trade Organisation and put in place the new multilateral trading system, they will be making not an end, but a beginning.

13

What was Wrong with Structural Adjustment:

In Defence of a Much-Maligned Strategy

After decades of stranded development theories, ideologies and paradigms, "structural adjustment", with its demands for clean fiscal policy and an end to uneconomic state enterprises, political privileges, market and exchange rate intervention and corruption, entered the aid arena like a refreshing dawn after a long night of frustrating dreams. Only the "old guard" of planned economy advocates and jealous academicians who had missed the boat were able to shut their eyes to the moral and economic justification of this liberating breakthrough international development policy spearheaded by the Breton Woods institutions then steered by some exceptionally courageous economists.

Reaction to SAPs

As with any revolution, defeat is awaiting the pioneers at the hands of political power greed, reactionary tactics by the formerly privileged and academic envy. The principal device serving the reactionary forces as a lever of influence on the mood of the "development community" has been the identification and dramatisation of new pockets or strata of (principally urban) poverty allegedly created by structural adjustment measures, while shunning the much broader-based rise in economic activity, real incomes and sense of fair reward in the overall society, especially the rural population.

That the hardship experienced by urban poor, formerly privileged under consumer price control and import subsidies to the debit depressed farm prices or maintained by grossly over expanded public payrolls, was only laying open the camouflaged erosion of the economy and near-bankruptcy of governments and public enterprises, was conveniently downplayed.

These reactionary howls were to be expected. Not that they met the entirely innocent. There had been naively sweeping, overly assuming demands by some structural adjustment missions. But an intellectually vigorous and dynamic "development community" would have coped with the ensuing opposition, strengthened the analytical and monitoring capacities and the political will to endure also rocky roads and bitter medicines on the way to a healthier base. Instead institutional rivalry, political opportunism and emotive populism were thriving. In a way, the "development community" behaved as if it did not want its patient to become able to stand on his own feet and eventually steal its raison d'être.

Worst, the Bretton Woods Institutions themselves, partly under the pressure of the emotive opposition described above fell to the temptation to rescue their lending volume, which was threatened by the frugality dictated to Third World public budgets under structural adjustment recipes, through hardship-easing loans. They thereby corrupted their creation in using it to reinforce their indispensability. As a consequence it soon turned out that some of the most obedient loan takers under structural adjustment terms experienced sharply rising indebtedness, exploited as a disqualifying symptom by the anti-structural adjustment camp

Whatever the opinions on structural adjustment policies, the commitment to the principles of "good governance" has come to stay, at least on paper, as an almost standard conditionally for official development aid from OECD donor countries. The realisation, matured in the implementation of structural adjustment programmes, that not the quantity of aid, but the quality of Third World government determines the positive or negative course of development, may be regarded as the most valuable fruit of the decades-old policy debate in the 'development community". And the use of aid a pressure or bribing factor towards "good governance" as foreign aid's least disputable purpose.

Out of the Limelight

Nothing, however, must be taken for granted. Achievement breeds its challenge! Structural adjustment, though in essence hardly disputable has been pushed out of the limelight and replaced by the oldest actor in the company, eradication of poverty, twinned with an equally perpetual endeavour at the macro-level: debt-forgiveness. This falling back to square one in donors approach to the problems of the south, i.e., the call to alleviate poverty and priorities direct efforts to this end above all other developmental efforts—does it indicate a sell-out of constructive ideas in the "development community"? Has any noteworthy progress been achieved in the past by this approach?

By telling a frugally toiling but independent subsistence farmer that internationally his condition is classed as "poverty", deserving compassion and support by the world community and cancellation of his debts, one can hardly expect a sustainable improvement in his output, satisfaction, or self-respect and even less, when he realises that the help principally provides jobs, fringe benefits and self-importance to a gamut of intermediaries, at home and abroad.

What do those poverty advocates (the"Lords of poverty") really know about the resources, life managements, value systems and ambitions of those they generalize by the billions? The great variance in the conception of life situations, from different external viewpoints.

What the aid system can do for these rural populations classed as "poor"/"underprivileged"/"exploited", is press for justice, i.e., "good governance". The achievement of structural adjustment policy through e.g. abolishing official price and exchange rate distortions, import subsidies and exploitative state agencies, has brought massive income improvement for peasant populations, i.e. the majority of LDC inhabitants, in dimensions unreachable by whatsoever direct "attack" on rural "poverty". What people want is not being benevolently treated as poor, but being justly rewarded for their work, i.e., by access to the unmanipulated market of their output. Slackening on structural adjustment/"good governance" conditionally under the present "10 year itch" for paradigm change means foregoing much of the potential opportunities for undoing injustice and exploitation of the masses. It should be clear where priority focus should be placed in ODA policy.

Small is Not Beautiful

The direct attack on "poverty", orchestrated by the Bretton Woods institutions under their freshly launched Poverty Reduction Strategy Paper (PRSP) campaign, is being rightly regarded as primarily an NGO domain, since most activities are expected to be carried out at local community level. This would require careful screening and coordinating of NGO activities and their integration via gradual expansion of their experience. But "small" is not "beautiful" for the development financing institution. Disbursement needs are pressing, calling for the new paradigm to quickly provide channels for another wave of loans to the "IDA Countries". Their problem of heavy indebtedness, which would principally exclude most of them from any new loan consideration, shall be solved with one stroke (which only the well-cushioned development bureaucracy can afford); debt relief against presentation of country PRSPs by the respective governments. NGOs are expected of play in the system especially the knowledge gap about the "poor" people's real wants and needs NGOs will naturally be tempted by such expansionary boost to their involvement (referred to sarcastically as their philanthropic empire" by an African conference participant), but this will not be conducive to quality and accountability of their performance, which ideally should be based on piivate sponsorship in combination with strong target-group provided self-help components.

Patience and Self-Restraint

Local knowledge and initiatives cannot be obtained under time pressure. "The grass does not grow faster by being pulled". When will the "development community" learn patience and self-restraint in the approach to LDC's capacity for constructive absorption of aid programmes accompanied by a genuine sense of ownership?

After all these deliberations, how shall development policy be shaped in order to better correspond with reality, without sinking deeper into hypocrisy and frustration?

To come back to the opening question: what was wrong with "structural adjustment"? Nothing was wrong with its intent. In fact this was very right and long overdue. Its implementation, however, lacked patience, perseverance and solid support from the development community, apart from its being corrupted as a vehicle for expansionary lending policy. If aid is meant to not be an end in itself, then structural adjustment policy needs constant reinforcement, underpinned by strict lending discipline. There should be an end to irresponsible lending and easy escape from its consequences by wholesome periodic debt relief burdened on the international tax-paying community. No ODA, either loans or grants, should be made available to governments who are not in active process of implementing "good governance" principles. A monitoring unit, reporting to the donor community on government performance in regard to its "-good government"? Structural adjustment commitment, should be maintained in each and receiving country by "donor consortia" comprising all locally represented bilateral and multilateral development organisations currently extending technical, financial or material assistance to the country.

In order to accommodate the poverty focus without diluting the necessary structural adjustment orientation of ODA, a division of activity-focus between the latter and the NGO sector would seem to be advantageous.

- ODA, limited to the countries abiding to structural adjustment/"good governance" conditionally, with focus concentration on sustainable physical, social and economic infrastructure principally at national and regional level, public management training, higher education and research, consultant and senior adviser services.

- The NGO sector, principally funded by private sponsorship, united to structural adjustment conditionally (but preferably grafted on local self-help initiative), with focus-concentration on the "Third World "poor", i.e. mostly at rural community and low-income township level, for amelioration of living conditions and local resource utilisation.
- Strengthening of linkages between the NGO sector and the UN Technical Agencies to mutual benefit: NGOs in need of professional information, evaluation and advice, of forum for discussion to find an actively supportive window at the agencies; the latter to maintain and develop field contact of research and policy generation, not least as a substitute for their declining project work (giving way to greater concentration on their global functions i.e., serving as information, policy initiation, and coordination/negotiation centre on topics of global concern such as e.g. : human rights, global monetary and trade systems, tropical forest and global marine resources, global and regional health threats, international standards).

In conclusion, it may be called to mind that aid and its institutions have no claim for permanence. They are justified only as temporary functions in a phasing out process of self-help support. Any claim for unlimited continuity would breed lasting infantilisation.

Add Value, go Global:

Can Southern Firms Break into Export Markets?

The global economy has changed beyond recognition over the last decade. Widespread economic policy reform and in particular trade liberalisation have open up new opportunities for developing countries. In poor countries, however, the consequences of trade liberalisation are not always positive. What can the private sector do to respond better and make the most of new trading opportunities? What factors have limited the impact of economic reforms on export performance?

Why have exports from poorer countries failed to increase more rapidly following trade liberalisation? What can be done to improve performance? Research on the response of firms in the private sector to economic reform can underpin new approaches to export promotion for poorer developing countries. For a long time, protective trade policies, poorly performing state-owned industries and state controls over the private sector were blamed for poor export performance in Africa and South Asia. Now that some of these problems have been remedied, other obstacles have come to light.

The effect of economic liberalisation and adjustment on the performance of poor countries has been cause for concern. Trade liberalisation should increase incentives to export and facilitate business enterprise by encouraging private ownership through privatisation and by attracting foreign investment. Macroeconomic stability ought to boost business confidence and performance. All these factors should promote exports, offsetting job and income losses caused by the closure or reorganisation of inefficient enterprises and industries yet, although some degree of reform and stability it is without export growth that was expected.

Trade reform and macroeconomic stability may be necessary conditions for improved export performance put by them are insufficient. The obstacles to improving export performance are numerous and there is no easy policy answer. The research programme examined export performance at three levels.

- Regional: how trade strategies should vary with skills and natural resource endowments
- National: factors influencing the export performance of manufacturing
- Sectoral: the performance of particular sectors of the economy.

The East Asian economies have shown that developing countries can complete successfully in global markets. For many, they provide a blueprint for economic growth applicable to many poor countries

South Asia's comparative advantage lies in its abundant unskilled labour, while Africa's lies in its abundant natural resources. Different export promotion strategies are essential. South Asia's best prospectus are in labour-intensive manufacturing: the region's low level of exports would soar over the next decade if current obstacles to trade were reduced. Africa's exports could also increase but its biggest potential in primary products that need little educated labour and abundant natural resources.

Some African countries could also be substantial exporters of manufacturers, but their actual manufactured exports in most cases now fall far short. Comparing Ghana to Mauritius—one of Africa's most successful exporters of manufactured goods differences in firm-level efficiency are apparent Mauritian firms have more capital per worker and use it more efficiently. Reducing trade barriers is not sufficient. Wages in Ghana would have to be substantially lower to offset low labour productivity. Alternatively, labour productivity will have to be drastically improved if Ghanian firms are to compete successfully in export markets with wages at current levels.

Even when companies use capital and labour efficiently, poor infrastructure is a frequent stumbling products to export markets—an acute problem in landlocked countries and equally acute for manufacturers as research on Uganda clearly shows. What huts manufacturing exporters is being hit by the high cost of transporting their output to foreign markets and of transporting the materials they need from abroad. The cost penalties resulting from geography and poor infrastructure are far greater in Uganda than from high tariffs and other import restrictions.

Southern firms can still break into export markets, however, developing—country firms do export to markets with exacting standards for product quality, reliability of delivery, and consumer safety. Two crucial aspects, however, are often overlooked:

- Non-manufacturing sectors, such as tourism and horticulture, generate significant employment and offer opportunities for supplying increasingly sophisticated products. Although manufacturing is considered more attractive, certain areas of tourism and horticulture can be equally appealing.
- New export opportunities are created as southern producers establish closer links with foreign customers. Producers of labour-intensive products such as garments, horticulture and footwear frequently depend on large retailers and specialist international traders for designs, information about demand and technical support.

Supermarkets make key decisions about which fruits and vegetables to grow, how they should be produced and processed and which firms should be included in the business.

Strategic decisions by international producers and retailers in the footwear industry have been crucial in developing new production locations such as Vietnam and Romania. Similarly, work on automotive components production in South Africa and India illustrates how global sourcing by the leading motor companies closes off some markets and opens up others. Export prospects can only be evaluated in the light of global restructuring in these industries.

Emphasising global linkages does not mean that developing countries are powerless in the face of global forces. Even in tightly-structured industries, there is scope for national policy and national strategy. Further more, there are important export sectors that are not structured in this way. Some tourism is dominated by large northern firms and is heavily import-dependent, but there is also enormous potential and national policy will be crucial in shaping the industry and its contribution to the economy as a whole.

For southern firms to break into export markets, certain issues must be addressed, especially in Africa. Some are recognised as important policy issues—investing in human capital and improving infrastructure for example. As one set of constraints are reduced—such as removing policy—induced distortions through trade liberalisation—another set takes precedence. In response to the integration of global markets, southern producers must join the global distribution chains to ensure markets for their exports.

These findings impose hard choice on developing countries. Should a firm allocate limited funds for investment in human capital or investment infrastructure? Future research might contribute by quantifying relative rates of return. On another level, countries may worry about the independence and autonomy of local producers if they are to join a global chain typically donated by northern companies. Rules regulate governmental trade and investment policies but who controls the global buyers and multinatinal companies whose decisions have such huge impacts on developing countries?

15

What's Driving Migration

The scale and diversity of today's migrations are beyond any previous experience. Rapid urban growth and environmental degradation in rural areas have led to internal migration affecting hundreds of millions of people. Migration is now seen as a priority issue equal in political weight to other major global challenges such as the environment, population growth and economic imbalances between regions.

Families and households form the basis for economic growth, social development and personal fulfilment. Decisions, by individual women and men on marriage, family, a place to live, shape the destinies of communities and nations. National policies and international conditions provide the context for individual decision-making. Effective development policies, including population, reproductive health and family planning policies, address this reality.

Data on national and global population trends set the agenda for national policy. An important element of population programmes is gathering data that will allow policy-making responsive to the realities of daily life, and to the needs and aspirations of individuals.

The dominant feature of global demographics is still growth. Age distribution is a growing concern, as the numbers of young and elderly people, grow, relative to the working-age population. The world is growing steadily more urban. From being a sign of strength and dynamism in the national economy, the rate and scale of urban growth has become increasingly a cause for concern. The influx of migrants to the biggest cities may be weakening both urban and rural sectors.

International migration is small in extent compared with internal movements, but has a disproportionate impact. Both internal and international migration are driven by population growth, and by inequities between countries. Migration is one of the choices which shape people's lives and the destiny of nations. But it can also be a symptom of inequity and underdevelopment. Migrants are by definition the most vulnerable members of the host community. Their living and working conditions should be protected.

Open and frank exchange of information and views between host and sending countries is needed more than ever. The aim of the international community should be to protect the right to move, but to ensure that movement is voluntary and that it stimulates rather than holds back personal and national development. "The point of departure should be the human right to live and work where one pleases, so long as it does not infringe on other people's rights to do the same."

The Urban Transformation

The rural sector is declining in importance and its contribution to national economies. It is increasingly part of a unified economy based on the city. Contact with the urban areas is easier than ever and is encouraged by rural development.

Temporary and circular migration is giving way to more permanent settlement. The largest cities are under increasing strain, and residents are encountering increasing difficulties in improving or even maintaining living conditions. Nevertheless, migration continues, driven by a variety of forces both positive and negative. The choice to move can be part of a strategy for survival or personal development; but it is often enforced by external conditions.

The urban transformation is irreversible, but the rural sectors must also be strengthened to balance the developing economy. Attention to gender issues will be crucial in ensuring a successful transition. The forces driving internal and international migration have much in common. Demographic pressures are contributing to both. As the pressures encouraging migration increase, the options for migrants become more limited. This collision is contributing to the atmosphere of crisis surrounding both urban and international migration.

Costs and Benefits

Migration is the result of individual or family decisions. But it is also part of social process. In economic terms, migration is as much a global phenomenon as trade in commodities or manufactured goods. It is part of a broader pattern, and evidence of changing economic, social and cultural relationships.

But migration may be evidence of a different kind of relationship; the combination of poverty, rapid population growth and environmental damage is a powerful destabilizing factor driving urban growth and eventually international migration. On the recipient side, migration has usually been seen as evidence of a thriving economy; today's industrial states were built in part by migrant labour, skills and investment. In today's increasingly uncertain conditions, migration may be seen as a threat to the security and well-being of the local workforce and society at large.

The only effective means to reduce migration pressures over the long term are to slow population growth; to stimulate economic growth and job creation at home, and promote the development of the individual and the family as the basic economic and social unit.

A Question of Gender

It is often assumed that most migrants are men, in reality, women make up nearly half of the international migrant population. Gender differences in social and economic roles affect migration decision making, household strategy, and the sex composition of labour migration. Attention to the gender dimension of migratory movements ought to be an important component in population and development planning.

Women frequently take the initiative in migration decisions, which may reflect limited opportunities in rural areas. Low status limits women's choices at home and may increase pressure to migrate, but it may also affect life in the host community. Opportunities may be limited by lack of education or skills, or by customer limitation on women's freedom of action outside the family or ethnic group. Paid employment for migrant women is usually in the lowest wage, least secure, and lowest status jobs, mostly in housework, child care and trade.

Most educated women end up in the same low-status, low-wage production and servicejobs as unskilled female migrants. Men too, experience downward mobility, but the contrast in the decline in women's employment status is far greater. Despite these disadvantages women migrants have become significant economic actors. Their status may be improved by migration, but the advantages are not clear-cut. Women's status as migrants is affected by their vulnerability, and by their lack of reproductive freedom. To ensure improved status they will need both legal protection and essential services, including reproductive health services.

Refugees

Refugees in the 1990s are overwhelmingly in Asia, Africa and Latin America. Their numbers are large, about 17 million, and growing rapidly. A further 3.5 to 4 million were thought to be in "refugee-like situation", though estimates are probably extremely conservative, and an estimated 23 million people internally displaced.

It is important to recognize the common roots of refugees and other forms of mass movement of populations. At the same time, despite the difficulty of distinguishing between political and socio-economic causes of migration, there is a clear need to distinguish between refugees and other groups of migrants. Participation in international efforts of burden-sharing would ensure that most refugee problems would be dealt with in their regions of origin.

Conclusions and Policies

Migration highlights linkages and interdependencies

within countries, with many implications for development agendas, including population programmes and development assistance.

Policies to regulate or moderate international migration have concentrated largely on urban growth. They have been only intermittently effective. The most successful have concentrated on stimulating rural development and the growth of alternative urban centres

Migration is also a personal or family decision, which is affected by external conditions such as poverty or environmental degradation, improving conditions of personal and family life can make a crucial difference in the decision to migrate, reducing dependence on migration as a strategy. Because migration is the result of personal and family decisions, it can be influenced by policies that improve the quality of life.

This offers the opportunity for policies emphasizing individual development, among them education, health (including reproductive health) and family planning. Such policies are particularly relevant to the strategies must take into account gender differences in social and economic life and the differential effects of policies.

Migration decisions are about family security and long-term-life-chances, rather than simply the maximisation of income. They are ultimately strategies designed to look after the individual's and the household's needs, safeguard their security, and respond to their aspirations. If the goal is to reduce migration pressures through development it will be essential to increase the capacity but reduce the need to migrate. Long-term external support will be required to make such policies a reality, particularly in areas of rapid population growth and potential mass outward flows. Highly co-ordinated allocation of development assistance can be help establish priorities and focus attention on basic needs. The challenge to both international donors and co-operating governments is to direct programme spending to the areas where it can be most effective.

16

Crisis and New Orientation of Development Policy

The poverty in the South, the dislocations in the East, and the orientation crisis in the North are not isolated phenomena. Rather, they represent an alarming amalgamation of dangers that are globally interlinked.

The low effectiveness of international economic and development policy is rooted in two outdated paradigms on which the present worldwide strategy of economic development is based, namely that:

1. The Western social and economic model optimizes the activation of productive forces—independent of the development stage of a country and its culture and therefore is best suited to satisfy basic needs.
2. It is possible to launch the development of a society from the outside within a few decades-without regard to its cultural and historical background—through external input of money, goods, technology, expertise, and personnel.

The twin paradigms of the timelessness and transferability combined with cultural ecological, and financial restrictions—have led international cooperation and development down the wrong path.

Only if we acknowledge the true dimensions of the global dangers, if we recognise the limitations and shortcomings of existing political instruments, and identify outdated theories and contradictory special interests, can we outline the cornerstones of a new policy of global cooperation.

Cornerstones of a New Development Policy

Starting with critical review of the shortcomings and paradigms of the prevailing development strategy, the following ten cornerstones of a new development policy are offered for discussion:

1. *Broaden the Concept of Development*

Whether a society is considered developed depends on the size of its percapita Gross National Product (GNP). Accordingly, the world is divided into a developed, semi-developed, and underdeveloped world. The yardstick for development, which has become the norm in the industrial countries, is one-dimensional: It only measures the monetary value of goods and services that are exchanged in the marketplace. This standard is too narrow economically because, it compresses the multitude and complexity of cultural, societal historical, social, and human values into a single economic category.

At the most, there can and should be agreement on what development and process should not bring about: Inability to find enough work to meet the most basic needs; exploitation and oppression of people; loss of cultural wealth and institutions; destruction of natural resources. These, however, are the very values that are scarified by the prevailing development strategy. In the future, development policy must do all it can to stop the loss of skills and self-reliance, the plunder of natural resources, the erosion of cultural values, the violation of human dignity and human rights. Initiatives must prevail which are orientated on these values, and not just on the GNP.

2. *Concentrate Development Strategy on the Internal Potential of Developing Countries*

There must be an end to the manic fixation of development strategy on external inputs and external markets. A new development policy must, above all, improve internal conditions for a productive economy, promote domestic production factors on a broad basis, protect cultural and natural resources, and greatly increase the domestic supply of basic goods. Wherever external inputs are unavoidable, credits must be strictly tied to the productivity and the ability of a country to absorb transfers. External transfers should be concentrated on "Software" for health, education, and social participation, administrative, and legal jurisdiction. Such an approach could also promote training and indigenous technologies, which are so important for economic development.

The set-up and expansion of the productive sectors must be decided, planned, and implemented by the developing countries themselves, and they must assume full responsibility. The external pressures, which force the developing countries into full integration with the world market, must be removed. This presupposes a structural reduction of interest rates.

3. *Make Development Policy a Central Feature of Polities*

Development policy must take the lead in mobilizing the various political forces and government departments to join the fight against the growing global dangers. It must ensure that the actions of all political departments are compatible with development policy is possible only if it becomes the central task of all political sectors, comparable to social and environmental policies, and the central goal of all policies. If development policy is to become a central task, development problems must become a priority in parliament and government. Society must understand that it is in the national interest to accept great global responsibilities.

4. *Reform the World Economy*

The industrial countries must abolish their protectionism in agriculture as well the processed goods sector. Simultaneously, the developing countries need to be protected selectively and for a limited time against imports from the industrial countries. The undifferentiated structural adjustment policies imposed by the IMF must be revised. The trend toward regionalisation of the world economy should not be opposed; rather, in the interest of both South and East, it must be regulated constructively to form a new, regionally based world trade structure.

A reform of the international finance system is urgently needed: Interest and exchange rates should not mirror the national interests of the big industrial states and the special interests of large banks and venture capital. Rather, they must reflect the global interest in monetary stability lower and stable interest rates, and sufficient development financing.

However, strengthening the international financial institutions is in the global interest only if the countries of the southern and eastern hemispheres are allowed to exert some influence. An international financial court must guarantee that violations of strict regulations to ensure international stability and solvency can be protested in a court of law.

5. *Redesign the Industrial Society*

As a global social and environmental policy, the new development policy must induce the industrial countries to give up their excessive consumption of air, water, soil, resources, and space. Increased utilisation of energy-conservation measures and environmentally friendly technologies is overdue. The economic and social policies of the industrial nations must promote balance rather than growth. This requires radical changes in traditional economic thinking, habits, structures and processes.

In view of limited world resources, unsatisfied existential needs in South and East, and continuous population growth in the south, the only premise for the future can be: Growth rates in the South must be higher than in the North, but they should no longer be in the North, but they should no longer be induced primarily by growth in the North. If economic policies continue to call for the North to provide the locomotive, the North will have to continue to acquire more resources than the south.

The North must relinquish the remaining growth frontiers to the South and East. The South must use this opportunity to activate its internal dynamic potential rather than integrate its economy with the North. However, ecological and social controls must be established at a much earlier stage than was the case in Europe.

6. *Strengthen Development Cooperation*

The share of official development assistance as a percentage of GNP, which dropped from 0.48 per cent in 1982 to 0.34 per cent in 1995 must be gradually raised again and reach at least 0.7 per cent in the year 2000—a goal which OECD established as early as two decades ago and which was reconfirmed at the Rio Earth Summit.

However, we must not succumb to the illusion that a doubling of ODA funds will even remotely meet the financial needs of South and East. State development policy must use it scarce public funds more effectively in the future. It must use restraint whenever partners in the developing countries can accomplish a task on their own and private initiatives and private enterprise are more competent to do the job. The government should be directly engaged only when it can be relatively more productive. Otherwise, it should limit itself to subsidizing private organisations.

7. *Now Orientation for Development Cooperation*

The state and its implementation agencies must abandon all direct responsibility for any projects which require unbureaucratic action, economic efficiency, and long term productivity. It must make a much greater effort to involve NGO's and private venture capital in development projects. At the same time, the state must insist and guarantee that private actions are compatible with social and ecological concerns.

In the future, the main thrust of government projects should be the promotion of the internal potential of a country. This comprises the political and administrative framework conditions of a humane, socially and ecologically sound development: Constitutional government, social institutions which facilitate broad participation of the population in politics, society, and economy; efficient savings, credit fiscal and financial systems; mechanisms for income, property, and land distribution which, promote productivity, justice, and social peace. In additional of this "software" of development, the following is needed: A regimen for the protection of resources and environment; measures to prevent the short term sellout of natural resources; elementary and general education and training, health care and social safety nets; capacities to develop science and technology.

8. *Reduce the Debt Service and Activate Private Capital*

Public funds must be used be to a greater degree for the financial rehabilitation of highly indebted countries in South and East; external demands for interest and principal payments must be adapted to the economic capacity of the respective country and its ability to execute external capital transfers.

Within the framework of international insolvency regulations, initiatives must be developed as condition for the continuance of the present rules for write-offs—which ensure effective cooperation from the banks and alleviate the heavy burden of private credits, with their high interest rates.

State development policy and private business interests should supplement each other. Government promotion of private enterprise initiatives for exports, investment, and employment in the developing countries must take into account their compatibility with development. In reverse, private engagements that effectively promote development must be actively supported by the government. A separate line item must be established in the development budget for such activation of private capital

9. *Set Regional Priorities*

State development cooperation has been scattering its scarce funds that not only among too many sectors, but also among too many partners. In the future, public funds must be concentrated regionally. More emphasis must be placed on regional programmes, and development cooperation with threshold countries must be enhanced. A portion of public funds should be set aside to provide an incentive for be set aside to provide an incentive for threshold countries to assist the poorer nations in their own region as well as deal with poverty in their own country.

The new development policy could then also help lessen ethnic-national conflicts and promote peace by sponsoring regional cooperation in joint development projects. For this purpose, regional development funds must be set up for cooperation in the transportation, energy, trade, and finance sectors and last, but not least for regional security systems and disarmament. Such regional funds could also provide the means to project refugees and improve their prospects for an eventual return to their homelands.

17

The End of the Old Order:

No Guarantee for Peace and Prosperity

In the last few years, the world has witnessed the collapse of communism and the end of many authoritarian regimes around the globe. In Eastern Europe and the former Soviet Union, multiparty systems with free elections have been introduced. In Afghanistan and Cambodia, Ethiopia and Angola, Nicaragua and Peru leftist government of various shades have given way to more pluralistic forms of government or are in the process of doing so. In south Africa, the white minority has accepted the black majority rules. All over the world, the old order which was established after World War II is breaking down. Freedom, Democracy, self-determination and economic prosperity are the slogans of the revolutions, which have swept the repressive regimes away.

But after the initial euphoria over the unexpected successes of the democracy movements, the world is now waking up to the sobering recognition that freedom does not automatically lead to peace and economic progress. In fact, in many countries that have managed to throw off the communist yoke ethnic rivalries have been sharpened by the call for self-determination. Age-old historical animosities between various nationalities and religious groups have come to the fore and are threatening to undo whatever gains have been achieved by introducing a pluralistic system of government.

The former Yugoslavia is a case in point. The state was an artificial creation born after the First World War and the demise of the Austrian-Hungarian and Ottoman Empires. Similar to many of the artificial states in post-colonial Africa, Yugoslavia consisted of a diverse mixture of ethnic and religious nationalities held together primarily by the charisma of former president Marshal Tito and the socialist ideology he imposed on the country. The wind of democratic change which blew across the European continent in the wake of Gorbachev's perestroyka also meant for Yugoslavia and hence it is divided between different ethnic or religious groupings.

What has happened in Yugoslavia is also taking place, although to a lesser extent, in Georgia, Armenia, Azerbeidjan, and Moldova where different nationalities have taken arms against each other to fight for their right of self-determination. In Ethiopia, the right to secession has been conceded to the Eritreans by the new democratic government, but already the Oromos and other ethnic groups are threatening to break away from the common state. In Afghanistan, the defeat of the communist regime has not brought peace to the country but the danger of a prolonged bloody civil war between the victorious guerrilla factions. Even in South Africa the end of apartheid marred by intensified inter-ethnic fighting between different African parties and groups.

The conclusion to be drawn from this review of recent developments around the world is certainly not that the call for freedom and democracy will inevitably lead to chaos. However, these developments must warn us that the wonderful concepts of Western philosophical thought will only work in practice if they are accompanied by the necessary spirit of compromise. Democracy functions well in societies, which are very homogenous, like many of the Western European ones. Even there the system may fail when faced with deep-rooted antagonisms as the case of Northern Ireland shows. But in countries which are divided by nationality, race, or religion freedom, democracy, and the right to self-determination may lead to even sharper conflict unless there is a genuine give-and-take between groups which grants every individual and every group the right to develop according to their own ideas. The older order is breaking down. The communist regimes have collapsed, most of the military dictatorships are on their way out. But in many parts of the world it is not yet clear what will come in their place. One thing is certain: there is no easy way to peace and economic prosperity. The new systems that emerge will have to guarantee the rule of law and the protection human rights. And they must try to instill a spirit of common values, which bridges the cleavages, which separate ethnic and religious groups. The oppressive regimes which held together their populations by force must be replaced by government which have learned the art of compromise—which is the essence of pluralism and democracy.

Employment and Promoting Ecology:

How a Service Culture Could Put People Back to Work

We are facing two big and urgent social problems: employment and Ecology. Both the unemployment of millions of people and the progressive destruction of the ecosphere are alarming. But they are linked with each other. The 'greening' of industrial products, processes and services could provide many more jobs.

Unemployment has many causes, including:

- Sluggish markets;
- Stagnating or declining purchasing power;
- Growing uncertainty about the future at all levels;
- lack of will and/or ability to innovate.

But joblessness is by far due mostly to the high efficiency of industrial machinery, which produces ever more, ever faster, with ever fewer workers.

Waste of Resources

The extremely high productive use of human labour and the extremely low productive use of resources are manifested by gigantic mountains of waste. Already today, the junked cars on scrap heaps alone would form a line that would reach to the moon. The scene is the same with discarded electrical and electronic appliances. Every year, millions of tons of ovens, washing machines, refrigerators, dishwashers, TV sets, entertainment electronics equipment and small appliances are being wasted.

If we throw away all these things after a relatively short time we are not only being wasteful and irresponsible with resources, but equally so with people's work. For with the products and materials we discard, we also dispose of the human labour they contain. It is imperative that we radically reduce the enormous turnovers of material and energy. In other words, the productivity of raw materials and energy must be markedly increased. Specifically, that means we must draw as many services as possible from one kilogram of material or 1 kWh of energy. Reducing the enormous flows of materials into the industrial system, as well as developing cycles of materials and responsibility (the manufacturer taken back and repairing and/or remanufacturing used products and materials) and the main pillars of a sustainable development that can cope with the future.

The industrialized nations must cut their consumption of raw materials by a factor of about 10 by 2050; if they are to be able to handle the challenges of the future. To achieve that reduction, innovation efforts must be directed at increasing resource productivity and/or ecological efficiency. In particular, strategies to extend the useful life of goods and intensify their use could result in reducing both the speed and volume of the flows of resources to industry.

Increasing Resource Productivity

In dealing with nature, we and industry are facing radical change. This is the transition from environmental protection (preservation of nature and health) to greater resource productivity (which at the same time means greater competitiveness). As a rule, environmental protection costs money, while higher resource productivity usually cuts manufacturing costs and/or increases a company's profitability. If the company can sell the same utility or benefits while using fewer resources, it saves twofold: in buying raw materials and on waste disposal. Thereby the rule is that goods and components cycles are more profitable than resources cycles, and that the company which is first in the market gains an additional competitive advantage in terms of a lead in knowledge and image. If a service, or benefits in the form of services, can be sold instead of products, the decoupling of company success and materials flows is even greater.

Impacts on Employment

The two social problem areas of work and ecology have to date been perceived and treated separately in politics, in industry and in our own minds. And, I believe, with little result. The link between the two must be established.

The strategies to boost resource productivity would have considerable impacts on the change in industrial structures, on handling existing product inventories, and on employment. In particular, the strategies would lead to a switch of focal point from a raw materials-intensive and use-value-related service economy. This is where another view of profitability comes in. Business management would no longer focus on value added, but on maintenance of value over longer periods based on the intrinsic value of a product. Expressed as a question, the value factor, which would move to the centre of business thinking and dealing, means: how can the utilisation value be improved and sold? How can products be made with as few raw materials and as little energy as possible and create a high benefit as pollutant-free as possible for as long as possible during their entire life-cycle?

With regard to employment, the production of long-life goods would appear at first sight to lead to a reduction in the need for work. In fact, however, the strategies to increase resources productivity have positive net employment impacts. The reason is that saving resources is based in principle on substituting energy by work, rather than the reverse as has been customary to date.

If the useful life of products is extended, that will not only preserve most of the materials and energy they contain as well as the work invested in them. The products will also require a considerable amount of mostly skilled work input. Reconditioning products is as a rule more labour-intensive than manufacturing them. So large-scale reconditioning and repair work increase the number of skilled jobs and at the same time reduces the inflows of materials and energy.

Comparing a car with a life-cycle of 20 years with two others that each have useful lives of 10 years gives a good example. The first car causes an increase in employment per life-year of about 50 per cent in terms of total work input in manufacture, service, repairs and reconditioning while at the same time reducing the energy consumption by half.

Regionalisation of Industry

Extending product service life would also mean replacing energy and/or capital by skilled work, helping to save money to boot. But not only rising costs of disposal, materials and energy would reduce consumption. Increasing transport costs would also mean that carrying all kinds of freight halfway around the world would make less and less business sense. That would result in ever more products and materials being circulated, reconditioned, and recycled or reduced on a regional basis. In turn, that would create regional jobs, and be more profitable as well as more productive of technology—not only from ecological aspects.

In addition, a way of doing business which encompassed material and responsibility cycles would no longer differentiate between manufacturing and reconditioning, or between marketing and remarketing. The structure of such an economy would be predominantly decentralised and regionalised so that it could adapt itself to the new cycles. It also would benefit from the greater efficiency of the new working practices.

True, jobs would be lost in the sectors of central production, and raw materials extraction and processing. But at the same time, more and higher-skilled jobs would emerge. These would not only be better qualified jobs, but also decentralized because reconditioning, repairs and maintenance must be done near the customer. And that, in turn, would also reduce goods traffic.

In addition, skilled workers would be needed because in many cases of small production runs it makes sense and is also more economical to hire such people. They can work faster and more flexibly—and mostly cheaper-than fully-automated production lines.

There also would be a growing need for maintenance, repairs and reconditioning. More and more people would be wanted for reconditioning, that is, the remanufacturing of old products. As reconditioning involves far more craft work than highly rationalised new production, there would be a positive impact on the labour market if there were more of the former and correspondingly less of the latter.

From Production to Services

Switching to long-life products and changing from selling products to selling use-values would strengthen the current trend of jobs shifting from industrial production to the service sector. For example, if the service of individual transport were to be sold instead of the product car, the company with the competitive advantage would be the one that had a service centre in every town and village, with appropriately staffed workshops and sales or rental facilities.

Enduring change towards a knowledge-intensive and use-value-related service economy would not only mean that more people would be needed to fill jobs. It would offer more opportunities for part-time work, as well as possibilities of employment for older people and the handicapped. People who earlier could not keep up with the pace of working life would be more inclined to return to it. Another impact would be that many companies would reduce their dependence on the world market. They would no longer switch certain tasks abroad, but assign them to their part-time employees, helping them to meet their commitments as self-employed entrepreneurs.

The latter would be accommodated by an ecology-driven fiscal reform which would make massive cuts or changes in subsidies and raise the cost of energy and raw materials consumption. This move would be accompanied by a reduction in income tax and non-wage costs such as social security contributions. The market would thus be more efficient, energy– and material-intensive new production more expensive, labour intensive repair work and reconditioning cheaper, and jobs would remain in the home country or region.

A number of more recent studies show clearly that an ecological tax reform would help to create jobs, and thereby could make a decisive contribution to reducing unemployment.

Give Developing Countries A More Favorable Deal:

An Assessment of the World Trade Conference in Doha

At the end of the 4th WTO Ministerial Conference in Doha, Qatar, the representatives of all WTO member states vigorously applauded Director-General Mike Moore when he dubbed the adopted work programmes for the new round of trade negotiations the "Doha development agenda."

The launching of a new round of trade negotiations with a broad agenda was the objective persistently pursued by the industrial countries, in particular the European Union, the United States, Canada and Japan. This objective has been achieved. Besides the continuation of the negotiations in the fields of agriculture and services, the Ministerial Declaration adopted by the Conference provides for the opening of negotiations in eleven additional fields. Undoubtedly a success for the industrial countries.

Clear Mandate for a New Development Round

The negotiating mandate, though, clearly reflects the political will to make the new round a "development round" with the aim of significantly improving the integration of the developing countries into the world trading system. To a large extent it takes into account the specific interests of the developing countries. Certainly a success with which the developing countries can credit themselves. A crucial factor of the course and the successful outcome of the Ministerial

Conference was, without doubt, the active involvement of the developing countries in the preparatory and negotiating process.

Doha Determines Merely the Work Programme for Negotiations

The Ministerial Declaration adopted at the conference merely determines the work programme for the new round of trade negotiations. Three factors contributed decisively to the positive outcome of the Ministerial Conference. There was a broad consensus among the WTO members states that (*i*) a second Seattle-like failure would put the WTO's workability at risk and was to be avoided at all costs (the 3rd WTO Ministerial Conference I Seattle in December 1999 ended in chaos without the adoption of a Ministerial Declaration); (*ii*.) the recessionary trends in the world economy were to be countered with the successful conclusion of the Ministerial Conference in Doha to improve the prospect for short-term recovery and, thereafter, sustained economic growth; (*iii*.) in response to the terrorist attacks of September 11, 2001, there should be a clear commitment to strengthen the rules-based multilateral trading system. Failure was, therefore, not an option, The strategic conclusion drawn from the Seattle failure was to limit the Doha Ministerial Declaration to establishing a broad, generally-worded negotiating mandate for a new round of trade talks that does not anticipate the outcome of the negotiations on controversial issues. The strategy worked. The deliberations at the Ministerial Conference focused on the scope of the negotiating mandate. The task of reconciling the conflicting interest between industrial and developing countries and working out a fair compromise has been left to the forthcoming negotiations.

Recognition of the Interests of the Developing Countries

In view of the objectives of creating a basis for sustained economic growth in the developing countries by better integrating them into the world economy and increasing their share in world trade, important preliminary decisions with regard to the forthcoming negotiations were taken by the Ministerial Conference:

- the Ministerial Declaration stresses the importance of implementing and interpreting the Agreement on Trade-Related Aspects of Intellectual Property Rights (TRIPS Agreement) in a manner supportive of public health and access to medicines; in recognition of the seriousness of the problem, a separate 'Declaration on the TRIPS Agreement and Public Health' was adopted; a number of public-health related issues have been referred to the Council for TRIPS for further deliberation;

- the Council for TRIPs has been asked to examine the relationship between *(i,)* the TRIPS Agreement and the Convention on Biological Diversity and *(ii,)* the protection of traditional knowledge, taking full account of the development dimensions;

- numerous problems regarding the implementation of WTO agreements are dealt with in a separate 'Decision on Implementation-Related Issues and Concerns' adopted by the Ministerial Conference; outstanding implementation issues are to be addressed as a matter of priority by the relevant WTO bodies;

- the Council for Trade in Goods will examine the proposal to bring forward the liberalisation of the textile sector under the Agreement on Textiles and Clothing;

- as regards agriculture, comprehensive negotiations were agreed on, aiming at: substantial improvements in market access; reductions of, with a view to phasing out, all forms of export subsidies; and substantial reductions in trade-distorting domestic support;

- as regards market access for non-agricultural goods, negotiations were agreed on, with the aim of reducing or, as appropriate, eliminating tariffs and non-tariff trade barriers in particular on products of export interest to developing countries;

- recognition of the principle of special and differential treatment of the developing countries as an integral part of all WTO agreements;
- technical cooperation and capacity building have been recognised in the Ministerial Declaration as 'core elements of the development dimension of the multilateral trading system' and firm commitments have been established in various paragraphs.

Turning the Ministerial Declaration's Spirit into Practical Policy

With these preliminary decisions regarding the agenda of the forthcoming negotiations, the course is set for the better integration of the developing countries into the world economy. To stay the course, there must be clear commitment and political will on the part of the industrial countries to make the new round a 'development round' by taking the developing countries' interest fully into account, being prepared to make meaningful concessions, and making good on the promise of significantly increased trade and investment-related technical assistance.

In the course of the negotiations it might prove a problem that many of the obligations in favour of the developing countries are formulated rather vaguely. The Ministerial Declaration is confined to declarations of intent even where—with a certain degree of goodwill—binding commitments would have been politically feasible. The bringing forward of the liberalisation of the textile sector, a key demand of the developing countries, has been referred to the Council for Trade in Goods for examination; this is certainly an expression of the industrial countries' willingness to compromise, but is in no way anticipates the final decision. As regards the objective of duty-free and quota-free access for all products of the least developed countries to the markets of the industrial countries the Ministerial Declaration simply repeats the commitment which was already expressed in the United Nations Millennium Declaration of September 2000, at the 3rd United Nations Conference on Least Developed Countries in Brussels in May 2001, and at the G7/8 Summit in Genoa in July 2001. Except for the European Union, no party has put this commitment into practice so far; the United States and Japan in particular have shown little enthusiasm for introducing duty-free and quota-free of all LDC products.

What makes us believe that the Doha Ministerial Declaration, will make a difference? The chapter on agriculture is more specific in that it provides for negotiations aimed at significantly improved market access, reductions/ phasing out of all forms of export subsidies, and substantial reductions in trade-distorting domestic support. However, a clear road map including a timetable for the negotiations and specific benchmarks for the reduction targets were beyond Doha's reach; in addition, the qualifier that the commitment to comprehensive negotiations does not prejudge the outcome of these negotiations leaves a back door open. In conclusion: If you remove the merely rhetorical phrases—such as "we place the developing countries' needs and interests at the heart of the World Programme adopted in this Declaration", "to take fully into account the development dimension...",— from the Ministerial Declaration, it becomes quite clear that the text contains relatively few 'programming elements' with a view to the development agenda of the forthcoming negotiations.

Fears that the vested interests of the industrial countries will re-gain precedence over development aspects in the course of the negotiating process are certainly not entirely baseless. The 'steel war' the United States is about to declare on the rest of the world clearly indicates that the Doha fair weather period is over. Business as usual has returned. The American steel tariff threats prompted EU Trade Commissioner Pascal Lamy to speak of a "perverse signal at a time when the ink is barely dry on the Doha Agreement."

The non-governmental organisations have a decisive role to play. It is their role to monitor the new round of trade negotiations, to make the negotiating process more transparent, to create public awareness with regard to the issues at stake, and to build up political pressure with the objective of making sure that development aspects are not pushed to one side and that the interest of the developing countries will makes their way into the agreements to be concluded.

Coherence of Trade Policy and Development Policy

The negotiating mandate for the new round of trade talks adopted in Doha has brought development politics onto the agenda of the WTO. The mention of development aspects in the WTO set of rules and regulations is not, in essence, new. In fact, the development dimension is recognized as an integral part of the general WTO mandate to foster economic growth. However, the particular importance the Doha Ministerial Declaration attaches to the consideration of development aspects in the negotiation process (it seeks, as it is put there, "to place the developing countries' needs and interests at the heart of the work programme") offers the opportunity to achieve greater coherence of trade policy and development policy. In this respect, the Doha Ministerial Declaration reflects the same trend as the "Everthing-but-Arms-Initiative" (EBA) of the European Union. Subsequent to its adoption by the EU member states, Pascal Lamy emphasized the coherence aspect as the characteristic feature of EBA Initiative (outweighing the shortcoming relating to bananas, rice, and sugar) by saying, "It is the first time that the European Union's trade policy has been substantially modified by the necessity of contributing to development policy." This perspective also characterized the 3rd United Nations Conference on Least Developed Countries in Brussels in May 2001.

To sum up, it can be said that the Doha conference has sent out an important signal for the process of coordinating trade and development policy with the long-term objective of achieving a coherent policy framework. The next step towards greater coherency can be taken at the International Conference on Financing for Development in Monterrey/ Mexico in March 2002.

Sustainable Development as the Guideline for Further Developing the Multilateral Trading System

The Ministerial Declaration reaffirms the commitment to the objective of sustainable development, as stated in the preamble to the Marrakesh Agreement of April 1994 (i.e. the Agreement establishing the WTO). However, theory and practice are far apart. The negotiating mandate for the new round is too cautious a step towards integrating environmental and social aspects into the WTO set of rules and regulations to be able to bridge that gap. Looking at the three pillar of the sustainable development concept—economic development, environmental protection, and social protection, in a nutshell the following can be said:

The negotiating mandate for the new round deserves good grades as far as the first pillar, economic development, is concerned. The course is set for better integration of the developing countries into the multilateral trading system, thus giving them the chance of actually benefiting from further trade liberalisation in the form of trade-induced economic growth. The inclusion of the so-called 'Singapore issues', investment and competition, offers the prospect of a medium to long-term improvement of the business and investment climate in the developing countries. As for environmental protection; negotiations on a (very) limited scale have been agreed on, the desirability of further negotiations will be examined. This is certainly not a big breakthrough, but a first step towards integrating ecological aspects into the trade rules. Disappointingly (but not surprisingly), social issues were not dealt with at the Doha Ministerial Conference. The developing countries' resistance to even discussing social issues, such as core labour standards, in the framework of the WTO could not be overcome; the issue was considered an absolute 'deal-breaker'.

Outlook

The developing countries' consent to the launching of a new round of trade talks cannot disguise the fact that there are still significant differences of opinion over a number of issues, including such key issues as agriculture, environment, investment and competition, and that there is a great deal of mistrust on the part of the developing countries. The one-day extension of the Ministerial Conference alone is proof of how difficult the process of reaching consensus on the launching of a new round of trade talks and its agenda had been. In order to successfully conclude the new round, the industrial countries have to deliver on their commitments, such as improving market access for goods of export interest to the developing countries and increasing their trade-related technical assistance.

The assurance of increased technical assistance was a major bargaining chip in getting the development countries' OK for the new round. If insufficient funds for technical assistance and capacity building measures are provided, it will most certainly diminish the chance of getting quick results. In a comment on the forth coming negotiations, the British Economist also highlighted the credibility aspect and the need for significant concessions, "Poor countries remain deeply suspicious of the rich world's commitment to truly freer trade. They bitterly remember the Uruguay Round, whose benefits went mostly to the rich. For the new talks to succeed, those suspicions must be proven wrong. Europe and America must quickly open up their markets for farm products and textiles. They must show that environmental concerns are not going to become a backdoor excuse for renewed protectionism. They must reform their oft-abused system of anti-dumping rules. And they must deliver on promises to beef up poorer countries' capacity to deal with the intricate procedures in the world trading system."

The Doha Ministerial Declaration offers the prospect of long-term gains for the developing countries. However, turning potential into actual gains requires tenacity in pursuing policies aimed at improving the business climate and, in general, the framework conditions for economic growth. Increased trade-related technical assistance and improved market access will not automatically result in growing export volumes for the developing countries. In addition, the strengthening and diversification of productive capacity is required. Successful integration into the global economy depends on tackling the supply-side constrains and other 'behind-the border impediments to trade' (ranging from weak infrastructure. Insufficient ancillary services and poor governance to macroeconomic instability). The Tanzanian Trade Minister, Iddi Simba, emphasized the complexity of the problems the developing countries are facing in his statement at the Ministerial Conference: "To operationalise the development agenda we need to have adequate capacity building which will go beyond addressing the normal WTO obligations. Adequate resources in the form of financial and technology transfer need to be in place to address the supply-side constraints. Along the same lines, WTO Director-General Mike Moore stated, "Capacity problems [in producing goods and services competitively], not trade barriers, are the major obstacles to growth in developing countries."

Concluding Remark

By creating a rules-based multilateral trading system, the WTO set of rules and regulations contributes to the shaping of the process of globalisation and to the emerging system of global governance. However, it can hardly be disputed that so far the industrial countries have been the main beneficiaries of the WTO-driven economic globalisation. We are still miles away from a true win-win situation. In a recent interview with the German weekly *Die Zeit,* the, Nigerian President, Olusegun Obasajo, criticized the industrial countries' hypocrisy, saying "Globalisation is a good thing. But only if there is a level playing field, from which all countries are able to benefit. You tell us that we have to open up our markets for your goods, whereas you keep your markets closed for our goods. Europe protects itself with innumerable trade barriers, everybody knows that. What kind rules are those?"

That is exactly what matters. The new round of trade negotiations launched in Doha must result in modified trade rules. Trade rules which take account of the specific economic constraints of the developing countries and are more favourable to them. The developing countries must be given the chance to 'cash in' on trade liberalisation and, strengthened by trade-induced economic growth, to pursue national pro-poor policies aimed at eradicating poverty.

E Learning-Designing Tomorrow's Education

The impact of information and communication technology is bringing gradual but often radical change to working, to learning and more generally to our way of life. Europe has to make the best use of the new technologies. This is particularly crucial in the education and training. Yet schools are often still inadequately equipped with computers and Internet connections. Technology-related skills among the staff concerned, in particularly teachers and trainer, need to be improved. There are significant divergences between countries and regions of this globe. In this situation future is at risk in the increasingly knowledge-driven global economy.

e-Learning

Our definition of 'e-learning' is 'the us of information and communication technology, including the Internet, to learn and teach. It also designates an initiative to:

- Encourage the development and the acquisition of digital literacy:
- Improve everyone's capacity to use new technologies for learning and working;
- Adapt our education and training systems to meet the challenges of the information society.

The general objective—to make Europe a major player in the 'new economy'—was agreed in March 2000 by the EU Heads of State of Government in Lisbon, e-learning falls under the European Commission's e-Europe action plan' and 'strategy for employment in the information society. These aim to bring Europe into the digital age.

The Objectives

- To generalize and improve access to hardware, software and to information and communication networks.
- To provide and simplify access to quality training for all of us.
- To develop cooperation between teachers, trainers and managers involved in establishing an educational area.
- To gather and share information on the best practices based on the use of information and communication technology for learning.
- To promote innovation,, know-how and expertise..

With Whom

The e-Learning initiative is open to cooperation with all interested parties and specialists in education and training, including;

- Experts in the use of technologies in the areas of education and training;
- Teachers, trainer and project leaders who develop innovative teaching practices in schools, universities and in all other educational establishments;
- Senior administrative staff in the education and training sectors;
- Pupils trainees, students, all those who want to learn with modern methods
- Information and communication technology companies, multimedia publishers, broadcasters;
- National, regional and local politicians interested in developing e-Learning initiative.

21

The dot.bomb Syndrome

Every one agrees that Privacy is good for business. But is there a market for companies offering to protect it? While several firms have recently suffered class-action lawsuits and public outrage over apparent intrusions into customers personal data, American Express in one of a number of businesses that may be blazing the right path from privacy to profits.

Last year, the credit card issuer rolled out "Private Payment", enabling its customers to go online and obtain "Disposable" credit card numbers that could be used for only online purchase if the number was ever stolen after being used for purchase, it would be worthless.

A key lesson is that protecting privacy, in this case the credit card number, became an "enhanced value" to an existing service. Moreover, American Express understood that the credit card number was the "tip of the privacy iceberg"—it plans to roll out a suite of related services, like anonymous internet browsing, in 2001. Rival companies like Visa/MasterCard, are following the same trail.

Another company to watch is PrivaSys, of San Francisco, which has patented technology to enable plastic credit cards to generate disposal numbers for each purchase PrivaSys does this by equipping the card with a calculator-styled key pad and LCD screen, a very thin battery, and a special magnetic stripe. The card holder punches a 4-digit PIN into the credit itself and—Voila!—the card generates the disposable number.

But securing payments is just one part of an emerging privacy protection market. Other firms are offering anonymous or "Pseudonymous" browsing tools, packages to control cookies—the strings of code that are planted on the user's computer by websites—and services to block hacker intrusion. Only time will tell which companies will control riche markets or achieve critical mass.

What's clear is the nascent e-commerce industry badly miscalculated the importance of privacy to their business model. For without privacy, there was no consumer confidence, and without such confidence, e-commerce, had no chance of fulfilling the short-term, high expectations that initially sent their stock values soaring. The current shake-out of the "dot.bomb" industry shows how costly this miscalculation over privacy was.

An Added but Essential Value

But if you though privacy was a big issue for e-commerce, consider the debate on the wireless industry. Referred to as Mobile or M-commerce, this industry appears to have unlimited potential to deliver information and location-based services, advertisements, and discounts coupons to cellular phones and hand-held wireless devices M-commerce risks hitting the major "Privacy button"; constant surveillance via location-based tracking; unsolicited ads clogging communications devices; detailed profiles of individuals interest and movements; and insecure payment mechanisms

Nonetheless, the industry is showing signs that it has learned from the mistake of e-commerce. A few months ago a workshop of the U.S. Federal Trade Commission, leading wireless industry groups unveiled guidelines requiring companies to obtain a consumer's consent before collecting or using personal data. This wasn't altruism: the industry understands that consumers will not tolerate commercial services to their cell phones unless they explicitly consent.

So far, privacy's brief tenure in the commercial market place indicates that some people are willing to pay extra to protect their privacy. But the more important lesson is that consumers are unlikely take more than a few extra steps to secure their data. If they perceive that a given medium, like the Internet is not privacy-friendly, they either change their habitual uses or actually refrain from them altogether. The likelihood is that privacy will become an "added value", creating a market for those who can integrate privacy into payment mechanisms without excessively burdening the consumer.

22

The Electronic Gap

In the United States, business journals, market gurus, economics professors and highly paid consultants talk incessantly about the coming global boom, the transformation of the workplace, the technology revolution and the knowledge explosion. It is implied that the world is slowly becoming a reproduction of Silicon Valley. It is asserted that this is the future. But instead of swallowing this hype, perhaps we should pull back and look at the globe as a whole.

The pessimistic view would be to point out what is currently occurring in Kosovo, West Africa, Rwanda, Chechnya, Kashmir and elsewhere. We might also follow Robert Kaplan in the trips he describes in *To the Ends of Earth*, only to discover that much of humanity is headed for disaster and self-destruction. I do not wish to be as negative as that. However, I would like to offer a caution to those who portrary globalisation in an uncritical and overly enthusiastic manner.

One in three Americans are regular, daily internet users. Even within American society, the computer and e-mail have widened the gap between educated people (chiefly whites and Asians) and the less educated (chiefly black Americans). This gap will be felt in every aspect of life, whether it is in opportunities, potential education or job-hunting. The United States will be divided into two groups, one which is computer-literate and the other which is not.

This phenomenon has been replicated at the international level. The most important fact is that we are in the midst of a technology revolution that sees less likely to close the gap between rich and poor countries than to wide the gap even further.

The technology revolution and the communications revolution still bypass billions of human beings The Internet may have more influence than any single medium upon global educational and cultural developments in the coming century. Yet only 2.4 per cent of the world's population is on the Internet, or one person out of 40. In Southeast Asia, only one person in 200 is linked to the Internet. In the Arab states, only one person in 500 has Internet access, while in Africa only one person in a 1,000 is an Internet user. This situation will not change as long as those lands lack electricity, telephone wires and infrastructure. They cannot afford either computers or the expensive software they require. If knowledge indeed equals power, the developing world may have less real power nowadays than it did 30 years ago, before the Internet was developed.

If we want to work toward a knowledge-based society in the coming century, over at least the next 10 years we need to make a concerted effort to bring poorer societies into the system of electronic communications. This effort will need to be coordinated by the World Bank, the UN Development Programme, UNESCO, the NGO community and the global business community.

The alternative is to perpetuate a world fundamentally undemocratic and structurally unsound. If we do nothing, if we let the knowledge explosion intensify in technology rich societies while poorer societies fall further behind, the growing gap between haves and havenots will lead to widespread discontent and threaten any prospect of global harmony and international understanding. This is the most significant challenge we face. We have no time to waste in responding.

23

When Computers Chip Away at Our Memories

Galloping advances in information technology promises to give us instant access to all worlds' knowledge. But how will human memory fare against the rise of the super-machine? If the architects of technology's next great leap forward are to be believed, all knowledge may soon be shrunk to vanishing point. Nanotechnology, or computing carried out at the scale of atoms, is their by word for the future. With its awesome potential, scientists have recently argued, around 11 million 400-page volumes could be stored and primed for instant viewing on device the size of a human palm.

The ink may still be fresh on these blueprints, but the elixir of portable omniscience no longer seems so far away. Seemingly cast-iron laws of ever increasing computer power, along with the rise of powerful new technologies, appear to point to horizon where all that can be known and remembered can be transferred to machines with which human beings then interact at will. And it is a future that for some is already spelling big trouble for the brain.

Surveys Point to Yawing Gaps in General Knowledge

Computers not only distract us from contemplation of deeper valleys; they discourage from contemplation itself. As surveys repeatedly show, knowledge of history, literature, geography and even current affairs seem to be on a steep decline: 60 per cent of adult Americans can not recall the name of the president who ordered the dropping of the first atomic bomb, just as 77 per cent of young Britons are perplexed by words Magna Carta. The day of the nano-shrunk library could soon come, but will any of its users be able to remember a single line of poetry?

The connection between these yawning gaps in general knowledge and the information technology is by no means established, but no means established, but a host of thinkers in different fields are sure the issues is one that will shortly become all to pertinent. At the same time as it help us and extends our physically capabilities, it diminishes our individual faculties. This is a vital question, one which has been around for a long time.

Good or evil, writing has nevertheless formed one of the main tools in the evolution of human memory. Indeed it is civilisation's unrelenting hunger for placing memory in external stores—cave paintings then manuscripts, libraries, printed works and finally computers—that has supported the entire march of the species. Each of these new technologies has helped humans "off-load" their memories Pre-literate societies, for instance, depended on oral tradition for their expertise—a practice undermined by the flaws of over worked brains, though fertile ground for epic poetry. Through the written word, memories were freed from the head: knowledge could be stored for retrieval in books, and then redrafted into the sort of novel and complex codes on which modern society is funded.

Becoming God Memory Managers

The benefits of storing memory outside the brain are unquestionable, but the invention of printing over 500 years ago followed by the post-war onset of computing have added a new note to the process: that of thundering acceleration. One simple equation has come to embody this. It stipulates that computing power—defined in terms of capacity and speed per unit cost—doubles every two years. The trend has held for the last 40 years. Should it continue as expected to around 2020, a personal computer by that year will have exactly the same processing power as a single human brain. Add the promised marvels of nanotechnology, optical and quantum computing, and machines might reach utterly daunting proportions. One penny's worth of computing circa 2099 will have a billion times greater computing capacity than all humans on Earth.

For many cognitive scientists, relations between mind and machine are already undergoing drastic reconfiguration. "Distributed intelligence" is the new maxim, encapsulating all systems in which individuals and computers mesh to carry out a collective task, whether it be landing an aircraft or tracking share prices. The Internet is so far the crowning glory—a system that in principle might combine individual users into a potent group min.

All of this may sound abstract, but the effects on memory are being felt now. Facts and figures no longer take pride of place in school curricula. Within the past two years, South Korea, Singapore and Hong Kong—heavens of rote learning—have debated plans to axe huge swathes of standard classroom study. Experts in education stress that students must learn to be adaptable, skilled in manipulating symbols, able to respond to new situations; in short, ready to deal with the new economy, a realm where the computer is king.

We will need a lot of new skills. We have to become good memory managers. We have moved away from managing a lot in our heads to managing memory devices. We have to devote more space to this executive control and less to rote memory storage.

Nurturing Imaginative Thinking at School

A heavy diet of ready-made computer images and programmed toys appears to stunt imaginative thinking. Teachers report that children in our electronic society are becoming alarmingly deficient in generation their own ideas and images. As Generics observes, human memory is much more than simple information processing. There are, for instance, at least five systems of human memory, making up an inordinately rich web of self-reflexive, interweaving recollection that no computer has even come close to imitating. But if memory is increasingly stored in machines that we then manage for our learning, work and leisure, then how will these systems in the brain fare? And how will imagination, intelligence and understanding—all of which depend on an efficiently functioning memory—be affected? The simple answer is: we still do not know.

Yet one image stalks the debate. It is not the old science fiction fear of malevolent computer but of a citizen without a personal memory to speak of, Bertman, for one, is convinced that boundless electronic information may be the deadliest enemy of human knowledge. It is no just enough to remember where we live, what our birthday is, and the name of our wife—there is more to human personality and identity than just the details we can find inside our wallet.

24

Fleeing the dot.com Era

The Internet has been proclaimed as the supreme network, the place where one day all human beings will communicate. So who are these strange people switching off in droves. Two years ago, one of the scientists behind the original computer language that gave birth to the internet announced a daring new plan. Work would begin, to spread the network beyond the boundaries of earth: the Inter Plan Net, as it has been baptized, would bring online computing to outer space.

Outlandish as it may at first seem, the plan is consistent with the hype and breezy forecasting all too often surrounding discussion of the internet. Even though mounting evidence points to a stubborn "digital divide" separating not just poor and rich countries but also groups within nations, policymakers, business and computer scientists downplay the figures as brief blips in the unstoppable onward advance of the network.

A figure of a hundred computers per human is not entirely unreasonable, leading to a thousand billion computers in the internet in 2020. This technological optimism has come under increasing scrutiny over the past year. The "Digital Divide", a UN Development Programme report revealed that less than 15 per cent of the world's population accounted for more than 88 per cent of the internet users. The financial failure o multiple dot.com firms and a sense that the technology is still too slow and unreliable have dented some grander hope for an "information superhighway." Yet it is a little known and barely noticed trend that may perhaps prove the most troubling: for the first time, evidence has emerged of a widespread tendency to abandon the internet.

No forecasting has yet taken account of this. A cast-iron premise in all extrapolations of the internet's future is that once the network is available, prolonged use necessarily follows.

Recent empirical work has suggested otherwise. Cyber dialogue, an internet research consultancy based in the U.S., has uncovered evidence of a slowdown in internet growth based in interviews with 1,000 users and 1,000 non-users. They argue that the rate of growth is decelerating overall, and that an absolute decline in the number of users aged 18 to 29 is underway. An interesting claim is that approximately one-third of U.S. adults simply do not believe they need the internet and what it offers.

The First Signs of a Rebellion Against Commercialisation?

Similarly, a 2000 survey in the U.K. found that 40 per cent of the adult population had no intention of going online, while most of them put their reluctance down either to cost or to the belief that the internet was irrelevant of their lives. Only a third of those who had stopped using the internet expected to reconnect in the future. Cyber dialogue estimated there were 9.4 million former users in the U.S., a figure that had jumped to 276.7 million.

People who stop using the internet, are poorer and less well-educated. Those who are introduced to it via family and friends are more likely to drop out than those who are self-taught or receive formal training. Most surprisingly, teenagers are more likely to give up than people over 20.

These results must be treated with considerable caution, since former users can easily become active ones at a later date. But the sheer existence of internet malcontents—and in such numbers in cyber dialogue is to be believed, raises critical challenges to the entire online industry. Why the decision to abandon use just at the time of giant internet expansion prompted by the introduction of the worldwide web.

Possibly the most salient feature in the trend is how it has coincided with a revolution in the network's character. Though there is as yet no concrete evidence of a link between the two, the suspicion is that the internet's first users have been alienated by the rampant commercialisation of the network over the past five years. Perhaps these figures point to the signs of rebellion against the transformation of a computer-linked community into a marketplace.

During its first 20 years, the network was the preserve of computer science professionals, students and academics resourceful enough to negotiate its complicated protocols. But in 1991, the U.S. national science foundation decided to allow commercial traffic, heralding the internet's switch from academic forum to virtual bazaar. The foundation's decision was followed by the creation of the commercial internet exchange, designed to regulate the exchange of traffic between newly emerging and profit-oriented internet service providers (ISPs)

From four computers connected in 1969, the number grew to 188 by 1979, 159,000 by 1989, and over 56 million worldwide by mid-1999. But the decisions of the early 1990s and the emergence of the web-allowing user-friendly multimedia features and full integration of the internet's previously dispersed elements (e-mail, file transfer and information access), also marked a change in style. The temptations of advertising on and profiting from this unique interface soon proved overwhelming. The proportion of internet computer addresses ending with .com or.net, and thus in the private sector, has risen steeply as a result. By 1999, these addresses formed 79 per cent of the internet, up from 47 per cent four year s earlier. Over the same period, the public sector share, represented by addresses ending with .edu,.mil,gov and.int, fell from 48 to 17 per cent. Non-profit organisations also dropped in their share, from five to two per cent. From their very low base at the end of the 1980s, business shot to a position of dominance within a decade.

There is no doubting that the internet remains an incredibly diverse collection of resources. Much of the original public sector ethos has also remained: most sites are free, and a slew of databases of archives held by companies (particularly the press) are open to whoever wishes t to access them.

But it does seem highly probable, if as yet empirically unproven, that the internet's mutation has driven away its first zealots. Back in the 1980s, there was relatively little distinction between the producers of what was on offer on the network and the consumers. If you had the skills and resources needed to access the internet, it was more than likely that you would be able to produce content and even services for it. The ethos was "give some, take some".

A String of Mergers in the Scramble for Supremacy

In the first few years of the web, this interactive culture continued, allowing anyone with a small amount of technical knowledge to become an "internet publisher". The change since then has been profound. Increased emphasis on scripting languages, multimedia and links from the web to databases held by organisations has handed power to skilled, professional programmers now, the medium that was initially conceived and an opportunity for smaller organisations and individuals to create and prosper on a "level playing field" has bifurcated into an industry of producers and a mass of consumers.

This trend was reinforced by the spread of search engines such as yahoo! and Alta Vista, both of which aimed to help the user navigate through the surfeit of internet resources. Commercial extension of this concept has given rise to the "portal"—one of most important elements in the current web industry. While offering traditional search features, these portals also try to maximize revenue by "click-through" advertising and deals with electronic commerce firms. The result of the scramble for portal supremacy has been a string of mergers and takeovers aimed at securing top spot in internet rankings, with leading contenders like AOL, yahoo! and Microsoft acquiring other firms, boosting content, and linking up to media companies such as Disney and ABC network. As the era of digital television dawns, this convergence of access and content is certain to accelerate.

There are definitely many users who would not object. Even in its mid-1990s incarnation, use of the internet was plagued by problems of connection, slow movement of data, excesses of information and the off-loading of "junk mail". It seems likely that some of those who abandoned the internet did so out of disappointment with the sluggish and haphazard service they received once they had signed up for the promised information revolution.

Such former users could well be tempted back by the latest internet developments. A stress on "pushing" predefined content through portals, sophisticated encryption schemes, secure payment systems and faster connections all promise to make the network easier to use and navigate, while simultaneously reducing variety. Add to this the fragmentation of the internet through its introduction in mobile phone, interactive television and palmtops, and it is clear that the marriage between access and content—between the device that leads the user into the network and what the user sees—may well become much tighter. The result could be absolute commercialisation.

The Symbolic Value of Joining the Information Highway

This goes to show that the requirements of different groups of users, and particularly groups of former users, are not necessarily the same. The more commercial the network becomes, the more it strays from its free and interactive origins, the more likely certain users will be turned of. A different group of users, on the other hand, might switch off were it to remain chaotic and technical

Certain trends may help to bridge this gap. The number and breadth of internet resources is still growing, as is the quantity of local content made outside the United States. Simple text message services sent via low-speed telecoms—the so-called "information dirt track"—could also speed the take-up of the internet throughout the developing world, evading the costly trap of commercial, high-speed service providers.

But a whole set of premises about the exponential increase in use of the internet and the loyalty of internet users are under threat. At present, the symbolic value of having internet access is mainly perceived as a sign of inclusion in an unspecified, high-technology future. Its rejection can be seen to imply a refusal of the intimidating, exhausting pace of technological and social innovation. The internet may travel to space, but it still has convincing to do on earth.

Labour Pains:

The Birth of a Movement

Cyber-rights and business groups are fighting a proposed cyber crime treaty. While there is strength in numbers, the groups' diversity may prove too much for the coalition to bear. An eclectic coalition of civil rights and corporate groups however, has launched an offensive against the proposed cyber crime treaty as well as a battery of controversial national laws and international standards.

Not surprisingly, the first privacy campaign emerged in the U.S. an Internet stronghold, against the infamous Clipper Chip. According to the government, this cryptographic device would have offered a standard for securing private voice communication. Two government agencies would have held the "keys", to be handed out only with "legal authorisation". Privacy-minded citizens quickly saw the dangers of this in light of the Federal Government's history of illegal domestic surveillance. In 1994, 50,000 people—a hefty chunk of the cybernaut population—signed the largest internet petition of its time against the proposal, which died soon after.

Perhaps the most striking feature in the battle to protect privacy has been the diversity of groups involved. The year 1996 saw a motley crew of immigration groups, gun-owners, liberals and conservatives band together to oppose legislation that would notably have extended wire tapping and allowed for more investigations of political groups. Irish-American and Arab-American associations joined out of concern that they would be more aggressively targeted in the "war on terrorism". Meanwhile, the fear of a more invasive government rallied gun-owners alongside groups from across the ideological spectrum. Once again, by tapping the Internet's power to organize and disseminate information, they shelved the legislation.

But this ad hoc co-operation is based on shaky ground. There was considerable coordination, for example, between civil liberties groups and the industry to oppose the 1994 Communications Assistance for Law Enforcement Act, which requires telecommunications carriers to modify their equipment, facilities and services in order to comply with authorized electronic surveillance. Yet once the industry received a promise of government funding to implement the law, it quickly abandoned the coalition. Corporate representatives then jumped sides again and sued the government over implementation rules in a controversy that is still brewing.

Old Enemies Become New Allies

With the Internet's growth, the privacy battle is becoming increasingly international. Most Western European countries have at least one cyber-rights group, a trend that is spreading across the continent and Asia, particularly in Japan. At the same time, existing human rights groups have also started to focus on the Internet. All it takes is for a single national government to ban free speech on the Web and the issue instantly takes on a global character.

For decades, international bodies like the Organisation for Economic Co-operation and Development (OECD), the Council of Europe and the European Union (EU) have been developing international standards relating to privacy, free speech and other civil liberties' issues. Their work has included brokering common rules on data protection and encryption policy to promote e-commerce. While some government representatives have put a stronger emphasis on protecting human rights, economic interests have clearly dominated the debate, strongly influenced by the International Chamber of Commerce, a powerful lobby of industry groups. Today, by pressuring governments bilaterally and multilaterally, the U.S. is leading the efforts to expand surveillance worldwide. This pressure amounts to what privacy advocates call "policy laundering": by pushing other governments into accepting controversial plans like the Clipper Chip, international standards will be developed which will in turn force the U.S. Congress to accept proposals it had originally rejected.

To respond with more muscle to these trends, a new opposition front emerged in 1996: the Global Internet Liberty Campaign (GILC), started by the Electronic Privacy Information Center, Human Rights Watch and the American Civil Liberties Association. The group now represents over 50 NGOs from some 30 countries. GILC operates by consensus. Member organisations propose specific actions such as drafting letters to world leaders, releasing reports and holding conferences. Member groups then agree to join in the action.

GILC and groups like the Trans-Atlantic Consumer Dialogue (TACD) are making inroads into the policy processes. Perhaps the most tangible signs of their success are the frequent invitations to participate in OECD meetings. But the movement has just one foot in the door: the next step lies in strengthening the role of NGOs outside the U.S. The problem lies in the old Achilles heel of international movements; a lack of funding.

26

Shhh... they're Listening

The journalist who first uncovered Echelon, a major electronic spy network, reveals how international surveillance touches us all. Constellations of giant golf balls can be spotted in the most remote locations across the world, from China's Pamir mountains to the swampy north coast of Australia and atop tiny coral atolls in the Indian Ocean. Between 30 and 50 metres wide, these smooth, symmetrical white domes also loom among rice fields in northern Japan and the vineyards and mountains of New Zealand's Island.

The clusters are the most visible signs of concealed electronic network that watch the world. Each dome is filled with satellite tracking dishes that silently soak up and examine millions of taxes, email messages, phone calls and computer data that keep business and political affairs afloat. Unknown to the communicators, their messages are running through the domes, into computer networks and onto listeners who may be on the other side of the planet.

As the world has globalized and international communications have become central to human affairs, these listening networks have grown exponentially. They are part of systems called signals intelligence or "sigint" operated by a handful of advanced countries.

For many years, sigint networks were secret: discussion of their existence was strongly discouraged or even forbidden by law in the countries concerned. Now the European Parliament is seriously investigating sigint organisations and their impact on human rights and international trade. Europe is focusing on "Echelon", a system that relies on listening stations in about ten countries to intercept and process international satellite communications. Echelon is just one part of an immense network run by the U.S. and its English-speaking allies—Britain, Canada, Australia and New Zealand—known as UKUSA after a secret agreement that created the alliance in 1948. Little escapes the UKUSA network, which intercepts messages from the Internet, undersea cables and radio transmissions as well as from monitoring equipment installed in embassies. It even operates in space with a fleet of orbiting satellites.

The history of systems like Echelon is as old as radio itself. The first international scandal over secret listening occurred in the 1920s, when the U.S. Senate discovered that British agents were copying every international telegram sent by American telegraph companies. Today's international networks were founded in the early years of the Cold War, when many western countries began jointly monitoring the former Soviet Union.

Fear not the word "Bomb"

Who is listened to, and why? Officially governments only admit that surveillance is aimed at commonly agreed perils such as arms proliferation, terrorism, drug trafficking the and organized crime. But this is the tip of the iceberg. The main aim is to spy on other governments' diplomatic messages and military plans while collecting information about trade. In fact, in 1992, the U.S. re-adjusted its national intelligence priorities, specifying that 40 per cent would be economic or "economic in nature".

While UKUSA is the world's largest network, France, Germany and the Russian Federation have similar systems. On a smaller scale, so do countries in Scandinavia and the Middle East, including Israel, Saudi Arabia and the Gulf states. The budges of all government sigint agencies probably add up to an annual expenditure of $20 billion, according to my calculations for a European Parliament report published last year.

Despite the extraordinary scale of Echelon and its sister systems, the press has mistakenly reported that the network could intercept "all email, telephone, and fax communications". Nor can it recognize the content of every telephone call. And it is pure fiction that by typing key words like "bomb" in an email, you can trigger a tape recorder in some secret base. For every million messages or phone calls intercepted, less than 10 will be used for intelligence purposes. Most personal communications are ignored except those of "important" individuals, like politicians, top business executives and their families.

The UKUSA network, however, does have the power to access and process most of the world's satellite communications and relay contents to client states. The system provides participating countries an enormous, unfair political advantage, since most developing nations cannot afford the expertise and equipment necessary to protect the privacy and security of their works.

Spying on the Government of the People

News about these systems began to leak out in the 1970s as U.S. intelligence agencies came under scrutiny in the "Watergate" affair, when former President Richard Nixon used electronic bugging against his election opponents. Since then an increasing number of whistleblowers have revealed the scale and effects of sign spying.

Over the next 20 years, official secrecy relaxed as U.S. Congressional investigations notably turned the spotlight on sigint agencies. In Britain in the 1980s, a controversial ban on trade union membership at the Government Communications Headquarters (GCHQ) boomeranged by shifting attention on its spying activities.

The growth of the public information culture on the Internet has taken these developments a step further. Now, even GCHQ and NSA have websites to reassure UKUSA citizens that they are not targets. No such safeguards apply to the rest of the world: these citizens are by default denied the right to privacy. The countries intercepting their communications are free to use the intelligence for whatever they wish.

Such conduct violates the Universal Declaration of Human Rights, and the International Telecommunications Convention, which assures the privacy of international communications. Indeed sigint agencies trample over a long line of treaties.

While individuals probably never know that they have been spied upon, their organisations and countries may pay a high price for it. During trade negotiations, sigint agencies can sweep up the messages of a producer nation to discover their bottom line. Armed with such secret reports, the negotiators for the developed world can force prices down to a minimum. Several governments have recently begun targeting environmental organisations or those protesting unfair world trade.

Even when there are no direct adverse consequences, the mere existence of powerful surveillance systems can exercise a chilling effect on free speech, inhibiting political and cultural development.

As these activities have become more controversial, the U.S. has tried to expand its circle of collaborators. Countries like Switzerland and Denmark are currently building new satellite station to gather and trade spy data with the U.S. But, as the ongoing European Parliament inquiry indicates, public awareness and concern are growing fast. Yet vigilance is not enough: if countries and peoples are to have equal rights in the global information infrastructure, concerted action must quickly follow.

27

Inclusion or Exclusion

Will the networked economy widen or narrow the gap between developing and industrialized countries? As we move from the industrial to the information age, access to the global information infrastructure for economies, companies and individuals becomes paramount.

The debate about the welfare implications of the information revolution for developing countries has given rise to diametrically opposed views. Some believe that information and communications technologies (ICT) can be mechanisms enabling developing countries to "leapfrog" stages of development. Others see the emerging global information infrastructure as contributing to even wider economic divergence between developing and industrialized countries. The reality is more complex.

A number of trends are broadly recognized as the hallmarks of the information age. First, ICT progress is expected to continue to promote the proliferation of communication networks as the costs of delivering these networks decline and the quality of their services improves.

Second, in a networked environment, the incentives for specialisation and outsourcing increase. This puts a premium on flexibility and responsiveness as business cycles shorten and interactions between producers and consumers expand.

Third, electronic commerce is expected to continue to expand rapidly and further contribute to the internationalization of service activities.

Fourth, information flows are at the very core of the globalisation process as countries and corporations project power by promoting their own culture and values on a global basis.

These trends suggest that the countries that are better positioned to thrive in the new economy are those that can rely on: widespread access to communication networks for their companies and citizens; the existence of educated labour-force and consumers; and the availability of institutions that promote knowledge creation and dissemination.

Income Inequality and Computers Literacy Gaps

The quality and coverage of schooling at all levels are also characterized by significant gaps between industrialized and developing countries. These gaps reinforce income inequality, not only internationally, but also within each nation as the ratio of female to male illiteracy tends to be higher the lower the level of economic development and the benefits of public education are typically skewed toward the richer classes. The gaps are even more dramatic when translated to the field of computer literacy.

Finally, developing countries are ill-equipped to implement pro-competitive regulatory regimes. In the same vein, the culture of protection and enforcement of intellectual property rights is often an alien concept. The same applies to reliance on networks to promote transparency and access to government services.

All these indicators seem to point towards a social transformation that will increase rather than diminish economic divergence at the international level. Developing economies would be condemned to fall further behind in the international economic race because of their lack of connectivity and ability to transform the information explosion into a knowledge revolution. Once one analyses the drivers of the information revolution, however, a different picture begins to emerge.

Technological developments are rapidly eroding economic and technical barriers of entry into communication networks.

Developing countries can, for example, leapfrog stages of development by investing into fully digitized networks rather than continuing to expand their outdated analog-based infrastructure.

First, by developing a modern information infrastructure countries can reduce isolation and exclusion. Many countries are experiencing fast expansion of cellular telephony as an alternative to inefficient conventional network services. Wireless technology can also provide affordable connectivity to rural areas in a fraction of the time that was required in the past to expand conventional telephone networks.

Second, countries can accelerate educational development by using their information infrastructures for distance education. The costs and effectiveness of such programmes are improving dramatically. ICT is also being used for lifelong learning, opening opportunities for new players in education systems. In developing economies, the dynamism of these new entrants can challenge conventional educational systems and play a catalytic role in their transformation.

Leapfrogging Stages of Development

Third, a modern information infrastructure can also be a powerful force for better governance. It can, for example, enhance tax administration, audition and control. Moreover, countries can now automate their institutions administering intellectual property rights, strengthening their efficiency and enforcement capability at a fraction of the costs that prevailed in the past.

In short, the logic of the networked economy is one of inclusion rather than one of exclusion. As technological progress continues to push the costs of computing and bandwidth down, opportunities for development-oriented applications of ICT will multiply. Moreover, for those already connected the value of the network increases exponentially as new participants join the community.

Technological Determinism and Government Policies

These considerations point toward a more optimistic scenario for developing countries' participation in the emerging knowledge economy. Although, no doubt, income and wealth inequality may increase in the initial stages of the process, catch-up can also happen at a much faster pace than in the past. ICT spending, for example, grew more quickly in most developing regions than is high-income economies in the 1992-97 period. And countries like South Africa and Brazil already boast a higher share of networked personal computers than most industrialized economies.

These scenarios can be criticized for sharing a common feature: technological determinism. The different outcomes predicted, however, illustrate that they are also influenced by other variables, in particular, government policies.

If, for example, developing countries maintain regulatory barriers to the expansion of networks, e.g., by favouring monopolistic providers for telecom services—the likelihood of the first scenario increases. In this case, global dualism will be magnified not only across the conventional North-South divide, but also in terms of country-level economic inequality as a small elite of connected people in the South benefit from the global information infrastructure.

On the other hand, if regulatory roadblocks are properly addressed and efforts to promote universal access to value-added networks and computers literacy are implemented, then the possibilities for catching up will proliferate. Participation in multilateral efforts, e.g. negotiations conducted under the World Trade Organisation and the World Intellectual property organisation—can also be used to leverage the process of institutional modernisation. Under these circumstances, the benefits of the revolution will be widely disseminated both at national and international levels.

The most likely outcome, however, is a combination of both scenarios in which a subset of developing countries is able to converge with high-income economies more quickly than ever before while others lag further behind. International efforts to promote pilot projects in this field can increase the number of countries in the first category.

28

Net Gains or Net Dream?

What is knowledge-intensive development? Can the Internet support it? Will this support make any difference to the lives of people in developing countries who are disadvantaged or marginalized? How can governments and other stakeholders ensure that their societies benefit from the new network technologies and services? This paper probes key implications of knowledge-intensive development and the internet. The new information networks can help, but governments and other stakeholders must introduce new policies to dismantle barriers.

Never before has so much chatter been heard in policy and business circles as well as in citizens' and ethical interest groups about the impact of advanced information and communication technologies (ICTS) on the global economy and the social order. For developing countries, these impacts are hard to track because of the rapidity of technical change and the unevenness of network and service availability. Education, training and skills development are failing to keep pace with the spread of the new ICTs, a growing source of anxiety for people in developing countries. Research is yielding contradictory evidence about who will benefit and how. The Global information infrastructure is penetrating the developing world and making claims on limited investment capital. Poverty, illiteracy, poor health, under-funded education, and worsening environment conditions are also making big claims on public resources. Everyone hopes that digital ICTs and the arrival of the vast networking potential of the internet will help remedy these ills.

The new ICTs are opening access to a flood of information from local and global sources. Converting this information into solutions to high priority development problems is the big challenge. The potential of ICTs must be harnessed to achieve major social and economic benefits. In principle, new digital networks and services could be used to communicate more quickly and cheaply, bringing the village to the world and the world to the village. The vision is one in which inclusion of people in developing countries in a more knowledge-intensive development process follows from access to services like the Internet.

'Net' visionaries portray a world in which access to the Internet and other information and communication services is all that matters. They acknowledge risks for people in developing countries, but they move on quickly to promote the use of new products and applications which, more often than not, have been designed in ignorance of development realities. The globalisation of markets for technologies and services, the rise of dominant firms like Microsoft, the emergence of a handful of global telecommunication operators, and increasing shortages of skills in key areas, mean that neither the benefits nor the risks of these technologies can simply be assumed. Harnessing networks and services to deliver benefits in developing countries means ensuring that those facilities are responsive to the poorest and most disadvantaged groups and communities. It means experimenting with new partnerships that boost equity, mobilise investment for building capacity and spur the harvesting and sharing of scientific and technical knowledge. It also often means pushing for national or regional participation in the global governance systems that steer trade, regulation, and intellectual property and privacy protection.

The new networks and services offer fresh opportunities for global and local change. But too much focus on the technical aspects of electronic commerce and new social applications means that organisational, social and cultural transformations are insufficiently heeded. Decision-makers in developing countries need to take measures to ensure that

ICTs can be used as 'tools' for social and economic development alike. Users need to accumulate new skills through formal and informal education, and learn how to use new sources of scientific and technical information to tackle problems creatively.

The articles that follow show that knowledge from ICTs can be converted into real social and economic benefits if new approaches like 'knowledge management' are effectively employed by governments and other stakeholders to overcome barriers. The Internet is enabling research results to be shared among community workers, businessmen and women, educators, and scientists worldwide. This exchange of knowledge is sparking novel ideas about how to harness ICTs to suit development purposes.

More research is urgently needed on the way ICTs are influencing the activities of women and men, on how skills and capabilities can be built up to tackle local and national problems, and on why some initiatives to use the new technologies and services succeed while others fail. It must draw upon the experiences of people in developing countries. Studies of global market trends and structures for ICT supply are needed, to gauge opportunities for people in developing countries to develop new services for strengthening their economies, creating jobs and reducing poverty.

Although there are substantial risks, potential gains from ICTs are far greater. Governments and other stakeholders should face up to the implications of the 'IT Revolution' National or regional ICT strategies should be set in place, corresponding to each country's development goals. They acknowledged that the next decades are not likely to see the gap between rich and poor vanish. But they argued that if governments and other stakeholders could find ways to use ICTs creatively, the gap at least could be reduced.

The social and economic exclusion of people in developing countries will not be eliminated by 'techno babble' about the Global Information Society, nor by dreams of 'Cybertopia'. However, when people's creative efforts and financial resources are combined to use ICTs to encourage innovative forms of knowledge-based development, there are likely to be substantial benefits. Action must be taken to ensure that visions beget policies so that ICTs bring more gains than losses to people in developing countries.

29

Wiring up the Ivory Towers

Prestigious universities are forging alliances to conquer a share of the e-learning market and stand up to virtual competitor. Just like airline companies, universities around the world are forming partnerships and consortia in response to the pressures of globalisation. The World Education Market held in Vancouver was a timely sign: the fair, expressly organized to foster relations between universities, training providers, software companies and representatives from nations with large education needs attracted participants from over 60 countries.

This race to "partner up" is fuelled by a number of factors. In most industrialized countries, government funding for higher education has decreased, forcing institutions to look for new markets either to subsidize campus programmes or just to remain viable. There is a growing need for lifelong learning as "jobs for life" vanish and the information society drastically reduces the shelf-life of almost any educational qualification. Technological developments, increasingly necessary for learners in all fields to master, offer ever more innovative tools for supporting e-learning.

For business, online learning is "the" new market opportunity with the need for re-training and professional updating predicted to crease an $11.5 billion industry by 2003. Business is better able to develop and maintain the technological infrastructure necessary to run large online systems and everyone, including the universities, recognizes that it takes robust telecommunications technology to deliver education and training on the scale demanded.

A host of companies has sprung up to help universities shape and package courses for online presentation, while network providers are jockeying of position to deliver online education.

The United States is the undisputed leader in the field, prompting governments in the U.K., Canada and Australia to commission being eroded by U.S. ventures turned global, Canada the U.K. are in the early stages of setting up their own virtual universities. But what has become clear is that the conservative and labyrinthine decision-making processes which characterize most university procedures are being jolted by a race to get a share of the lifelong learning market.

So far, the most common approach for universities to break into the e-learning universe has been to develop courses specifically for a corporate partner or to form alliance among themselves. Universitas 21, a company incorporated in the U.K. is a network of 18 leading universities in ten countries.

Very often, prestigious universities has stayed clear of going fully online, seeing a danger to their brand name. Many are limiting their offerings to continuing education programmes and/or non-degree courses, and more often than not, they are aiming at the corporate market. One Company UNext.com, has partnered with first-class institutions such as the University of Columbia (U.S.) and the London School of Economics to create online courses marketed under the name Cardean University. Their target: the Fortune 500 companies as well as individual adults. They've managed to attract Nobel laureates to design courses and the universities have formed spin-off for-profit companies specifically to develop online programmes. This facilitates the commercialisation of software and other products, and is a way to take a commercial approach to continuing and professional studies without compromising the University's standing.

Then there are the freestanding for-profit virtual universities which are arousing the ire of institutions that have prided themselves on a long history of public service.

The most quoted exemple is Phoenix University, the largest private outfit in the U.S. Now owned by the Apollo Group, it operates the country's largest online programme with 12,200 students. The university tracks students progress and contacts those who don't submit assignments on time or fail to enroll in subsequent courses. Many critics question Phoenix's blatant commercialisation, but few doubt the university's impact on continuing professional development provision.

Although e-learning is in its infancy, its impact can already be gauged. New providers are coming on the market all the time and the trend is accelerating to the point of upsetting universities virtual monopoly in educational accreditation. An Information Technology training course offered or accredited by Microsoft has undoubtedly become more valuable than a Bachelor of Science from a renowned university.

The more consumerist the approach of the education provider, the more what is taught is influenced by demand. MBAs dominate e-learning provision and IT courses are a close second. While the new consumer/learner demands flexibility, choice and just-in-time learning opportunities, suppliers will inevitably arise who are focused on meeting the demand at the expense of quality and value. And is the consumer really the best judge of what course material to choose? Education is a more complex "product" than toothpaste or washing powder. A totally consumer driven education market is unlikely to be in society's best interest in the long term. The commercialisation of education usually goes hand-in-hand with desegregation: course design, delivery, tutoring assessment and accreditation may be carried out by different organisations. Students might study courses or modules from different universities or providers and then put themselves forward for examination and accreditation by yet another institution. While most academics loathe marking assignments, they regard this scenario with horror, and blame commercialisation for the demise of the 'community of scholar's concept of a university. The death of the 'course' has also been predicted, with learners—especially corporate and on-the job learners—demanding short study modules. What then happens to the ability to get an overview of a field when learning consists of the students selecting a whole series of unconnected learning "bites"? Learners will be "zapping" between short sequences or presentations much as they do between television channels.

But while some faculty view e-learning with alarm, technology-based learning is where most of the pedagogical innovation is taking place in universities. Multimedia learning resources and interactive simulations are being developed for the web. Collaborative learning activities, new forms of online assessment and small group teaching technologies are making online courses more stimulating, interactive and attractive then many face-to-face taught courses.

Despite "doom and gloom scenarios", most moderate observers of the scene see a continued future for the campus university, especially at the undergraduate level, while e-learning will above all cater to adult professional and independent learners. Some commercialisation of education is good if it foster innovation, concern for quality and responsiveness to consumer demands. But if some is good, more is not necessarily better! Not in education at least.

30

Information Technology Outsourcing Goes Global

Many information technology jobs have been shifted to lower-cost countries, and may soon migrate onwards to regions offering even cheaper labour by affecting US job market. A recent study from the analysts; Forrester Research gives some credence to this concern.

The study suggests that as many as 3.3 million white-collar jobs, representing US $136 billion in wages, could shift from the US to lower-cost countries by 2015. Some companies embracing so-called offshore outsourcing believe they can get better quality work at half the cost. It is a similar story in the United Kingdom, where the telecomes union, CWU, has strongly criticized British Telecom (BT) for plans, announced in March, to transfer 2,700 UK jobs to India this year. The jobs identified include directory enquiries, billing videoconferencing and some telemarketing. CWU claims all BT divisions are exploring the possibility of transferring work to India and that thousands more jobs could be involved.

In its attack on BT, the CWU made it clear that it has no issue with India or with Indian workers. Its main argument is that, as a company which derives the bulk of its profits from UK customers, BT should have obligations to support the British economy by employing local workers. The union also warns that, by moving jobs to India, BTG would be setting a very dangerous trend, which could see hundreds of thousands of British jobs potentially at risk.

How New a Trend

The transfer of data processing and data imputting from developed to developing countries dates back to the 1970s. Then, Caribbean islands such as Jamaica and Barbados led the development of offshore" working. Initially, work typically consisted of routine administrative work, such as processing airline ticket stubs and credit card applications. Since then, however, much has changed; a sign perhaps of the apparent rise of the knowledge economy even in less-developed countries.

India, in particular, has developed rapidly in recent years as a destination for software programming and IT-related work. Compound annual growth of more than 50 per cent had pushed the industry from a value of about US $175 million in 1989-90 to some US $5.7 billion ten years later. The Indian trade association, NASSCOM (National Association of Software and Service Companies), predicts the sector's turnover could reach US $85 billion by 2008.

NASSCOM has identified "IT-enabled" services as the focus for major expansion in the coming years. Such services include IT management (for example, network management and maintenance), payroll processing financial services and client management, including order processing and call centre operations. In practice, the last of these-call centre operation—has so far been the most important part of India's rapidly growing IT service market.

Call centres employ some 100,000 mostly young people in India. They are highly educated, usually holding degrees in engineering or computer science, and their working conditions are good–typically purpose-built blocks in IT parks outside cities such as Bangalore and Mumbai. Call centre workers are trained to be effective when talking to customer's abroad, so that staff dealing with the United States will be tutored to speak with a US accent and to understand US culture. Those handling calls from Britain are similarly trained on aspects of British culture, including the British weather. In some instances, staff are encouraged to take on US or British names when talking to clients abroad, rather than using their own.

The main attraction of offshore outsourcing is the lower costs. Even though relatively well-paid by Indian standards, personnel costs are a fraction of Western wages. One British press report in 2001 suggested that Indian call centre workers earned the equivalent of $3,800 compared with UK workers' starting salaries of $19,000. More recently, the CEO of an Indian call centre company estimated that, in total, costs could be reduced by 40 to 60 per cent by moving to India.

Such developments are not limited to the English-speaking world. French companies are looking to African francophone countries, such as Mauritius and Morocco, as suitable destinations for call centres. Latin America provides an obvious location for Spanish companies looking to more offshore, while even German-language call centres have been established in lower-cost locations in developing countries.

This trend resembles the global relocation of work, which took place a generation ago in the manufacturing sector. What is new is not simply that this trend is now being extended into the service sector, but also that middle-class and professional workers in developed countries are for the first time becoming directly affected. White-collar as well as blue-collar jobs are potentially migrating.

Trade unions in developed countries understandably cite the danger of "social dumping" and a "race to the bottom". On the other hand, the development of indigenous software sectors in countries such as India does offer work opportunities for well-educated people to find work at home, as an alternative to seeking out work in North America or Europe—a practice sometimes derogatorily known as "body shopping".

North to South, or South to North

But the migration of jobs from north to south is only half of the picture. The other is south to north migration of workers (a familiar feature in the IT sector, at least uptil the bursting of the dot.com bubble). Naturally enough, developing countries which have invested in educating young people are becoming increasingly concerned that such education is being used to seek work abroad.

Still, there is real concern among the unions that established levels of social protection and labour standards could be lost when jobs migrate to lower-cost regions of the world. Perhaps paradoxically, the Indian IT sector itself may have grounds to fear the same thing. The issue is that other countries are prepared to compete to provide ever-cheaper locations for work. A recent report on global outsourcing explored the potential of destinations such as Malaysia, Thailand, Vietnam, Mexico and Brazil, as well as Eastern European countries with established high-technology sectors, such as the Russian Federation, Ukraine and Bulgaria. It is however, China, which is seen as the most likely next big player.

One trade union response to the fear of social dumping is to reassert their demand for the inclusion of core labour standards in future trade agreements negotiated through the World Trade Organisation. But there is another response happening at both ends of the outsourcing chain. This is the development of moves by unions to organize previously unorganized workers.

In the United States, people have been trying to persuade professionals in what has traditionally been a highly individualistic work culture to consider bargaining collectively with employers. "For many contractors and permatemps (agency workers with long-term employment with a single company), the prospect of obtaining that elusive permanent job, combined with fear of that never happening if they rock the boat in way, often outweighs any motivation to mobilize around key workplace issues. Yet, for many, that permanent job never appears. Meanwhile in India, fledgling IT Professionals forums are developing in the states of Karnataka (focused on the cities of Bangalore and Mysore) and Andhra Pradesh (focused on the state capital of Hyderabad). First established in 2000, the Forums have chosen not to use the term "trade union", which they claim has negative connotations among their target membership. However, the organisation have affiliated to the global union federation, Union Network International (UNI). Forum members say that, by working together collectively, they can better guard against professional risks and advance their careers. The forums mission statement is to become "the voice of IT professionals, to enrich and empower their knowledge, to promote their interests, and to contribute to the overall growth of the ICT (information and communications technologies) sector."

Enriching individuals knowledge may ultimately be one of the best protection measures against the risks of job migration, whether the flow is from the West towards India or from India towards other countries such as China. As with manufacturing a generation ago, the low-skilled, low value-added jobs will tend to be most mobile in a globalized world economy.

Bibliography

Frank Andre Gunder: "India in the World Economy, 1400-1750", *Economic and Political Weekly,* July 27, 1999.

Swamy, Subramanian: "Response to Economic Challenges: The Economic History of China and India (1870-1950)", *Quarterly Journal of Economics* (1979).

A.K. Sen: "Pattern of British Enterprise in India: 1854-1914" in *Social and Economic Development,* B. Singh and V.B. Singh, (eds), New Delhi, 1965, p. 420

Perkins, Dwight H. (Ed): *China's Modern Development in Historical Perspective,* Standford, 1975.

Blyn, George: Agricultural Trends in India-1891-1947 (University of Pennsylvania Press, 1966).

Raychaudhari, T: "The Mid-Eighteenth Century Background" in *The Cambridge Economic History of India,* Vol. II, Cambridge University Press, 1982 (Henceforth: CEHI)

Subramanian, S.: A Statistical Summary of the Social and Economic Trends in India (In the Inter-War Period), Office of the Economic Adviser, Government of India, New Delhi, 1945, Table VI.

Stokes, Eric: *English Utilitarian and India* (Clarendon Press, 1959), p. 134.

Brodkin, E.I.: "Proprietary Mutations and the Mutiny in Rohilkhand", *Journal of Asian Studies,* XXVIII, No. 4, August 1969, p. 667.

Cohn, Bernard: "The Initial British Impact on India," *Journal of Asian Studies,* XIV, No. 4, August 1960, p. 418.

Goldsmith, Raymond: *The Financial Development of India: 1860-1977,* Oxford University Press, 1983, p. 47.

Kuhn, Phillip: "Local Taxation and Finance in Republican China" in Jones, Susan (ed): *Select Papers from the Centre for Far Eastern Studies,* The University of Chicago [1972].

Cottrel, P.L.: British *Overseas Investment in the Nineteenth Century,*Macmillan, 1975.

Ashton, B., Hill, K., Piazza, A. and R. Zeita: "Famine in China, 1958-61", *Population and Development Review,* 10, 1985, pp. 613-645.

Fairbank, John: *China,* Harvard University Press, 1963.

Murphy, Rhoads; *The Outsiders: Western Experience in India and China,* (Univ. of Michigan Press, 1977).

In 1835, Lord Macaulay prepared a *Minute on Education,* which became the basis for English language education in India. of *Journal of Asian Studies,* XVII, No. 4, August 1958, p. 570.

See Report of the Destruction of Industries in North China, Chinese Delegation to the United Nations, 1948, New York.

For estimates, see Rostow, W.W: *Prospects for Communist China* Wiley, New York, 1954.

Domar, Evsey: *The Theory of Economic Growth*, MIT Press, 1966.

Swamy, Subramanian: *Eonomic Growth in China and India: A Comparative Appraisal-1952-70,* University of Chicago Press. 1973.

For details see Swamy, Subramanian: "Structural Changes and the Distribution of Income by Size: The Case of India", *Review of Income and Wealth*, June 1967.

Swamy Subramanian: "Spectral Analysis of Indian Prices: 1860-1967" in Bhatt, Mahesh and Mukund Trivedi (eds), *Liberalism and Less Developed Countries,* Gujarat University.

Swamy, Subramanian: "Investement Policy in the Indian Economy", Paper Presented at the University of Bolognia, Italy, September 1992.

Swamy Subramanian: "Government-Academia Interface", Paper presented at the Prime Minister's Consultative Conference at Namibia, February 26, 1997.

Das, Parekh and Parekh: *India Development Report* (1999), Oxford University Press.

Jayal, Niraja Gopal: "The Governance Agenda",. *Economic and Political Weekly,* Feb. 22, 1997.

Swamy, Subramanian: *Economic Growth in China and India [1870-1986]: A Comparison in Perspective,* UBS, New Delhi, 1989.

Reich, Robert (ed.): The Power of Public Ideas, Ballinger, Cambridge, and Mass USA, 1988.

Index

A

Africa, 35
African countries, 56
Agreement on Trade-Related Aspects of Intellectual Property Rights (TRIPS Agreement), 82
American telegraph companies, 110
Andhra Pradesh, 128
Antidemocratic process, 24
Anti-globalist, 1, 2
ASEAN countries, 3
Asia, 35
Asian economic problems of 1997, 6
Australia, 110

B

Bangalore, 128
Banking services, 6
Barber, Benjamin, 12
Berlin Wall, 12, 38
Bretton Woods Institutions, 50
Britain, 110
British Telecom (BT), 125

C

Canada, 110
Cash-strapped authorities, 9
China's Pamir mountains, 109
Cold War, 110
Common wealth, 6
Computers chip away, 96-99
  becoming god memory managers, 97-98
  nurturing imaginative thinking at school, 98-99
  surveys point to yawing gaps in general knowledge, 96-97
Council for trade in goods, 82
Crisis and new orientation of development policy, 64-70
  broaden the concept of development, 65
  concentrate development strategy, 65-66
  cornerstones of a new development policy, 65-70
  make development policy a central feature of polities, 66
  now orientation for development coopera-tion, 68-69
  redesign the industrial society, 67-68
  reduce the debt service and activate private capital, 69
  reform the world economy, 66-67

set regional priorities, 69-70
strengthen development cooperation, 68
CWU, 125
Cybertopia, 119

D

Defence of a much-maligned strategy, 49-54
development community, 49
ODA, 53
out of the limelight, 51-52
patience and self-restraint, 52-54
reactions to SAPs, 49-51
small is not beautiful, 52
Democracy and the market economy, 32-37
countries in transition, 36
democracy and economic freedom, 34-35
democratic government no guarantee for equality, 32-34
priority for growth, 37
Developing countries, 80-88
clear mandate for a new development round, 80-81
coherence of trade policy and development policy, 85
concluding remark, 88
Doha determines merely the work programme for negotiations, 81
e-learning, 89-90
objectives, 90
outlook, 86-88
recognition of the interests of the developing countries, 81-83
sustainable development, 85-86
turning the ministerial declaration's spirit into practical policy, 83-85
with whom, 90
Die Zeit, 88
Dispute over globalisation, 24
Doha Ministerial Conference, 84, 86
Dot.bomb syndrome, 91-93
added but essential value, 92-93
disposable credit card numbers, 91
e-commerce, 92
privacy, 91, 92
private payment, 91
pseudonymous browsing tools, 92
rival companies, 91
Dot.com era, 100-105
computer science professionals, 102
cyber dialogue, 101
digital divide, 100
first signs of rebellion against commercialisation, 101-103
internet first users, 102
– research consultancy, 101
multiple dot.com firms, 100
string of mergers in the scramble for supremacy, 103-104
symbolic value of joining the information highway, 104-105
U.S. adults, 101

U.S. national science foundation, 102
web-allowing user-friendly multimedia features, 102
Duty-free and quota-free of all LDC, 84

E

E learning-designing, 89-90
Echelon, 109, 110
Economic development, 86
– globalisation, 15
– liberalisation, 55
E-learning, 121
Electronic gap, 94-95
Chechnya, 94
communications revolution, 95
economics professors, 94
internet users, 94
Kashmir, 94
Kosovo, 94
market gurus, 94
NGO, 95
Robert Kaplan, 94
Rwanda, 94
technology revolution, 95
UN Development Programme, 95
West Africa, 94
World Bank, 95
Electronic spy network, 109
Employment and promoting ecology, 74-79
form production to services, 78-79
impacts on employment, 76-77
increasing resources productivity, 75-76
industrialized nations, 75
regionalisation of industry, 77-78
waste of resources, 74-75
English-speaking allies, 110
Environmental policy-making, 47
European Parliament, 109

F

Far-Eastern Four, 3
Financial and economic crises, 2
Free trade as peacemaker, 26-31
danger of a slowed-down world integration, 29-30
do not exclude poor countries from the competition, 30-31
exclusion as penalty, 28-29
free trade oppsition cloaks itself in dumping charges, 26-27
revolutionary success of open-market policies, 27-28
Freeing of trade, 1

G

G7, 48
GATT, 22, 27-28, 41, 46, 47
GDP growth rates, 28
Gender differences, 62
Ghanian firms, 57
Global information infrastructure, 117
Global Information Society, 119
Global linkages, 58
Global Village, 8
Globalisation, 1-5, 22, 24, 26, 29, 31, 88
– and knowledge divide, 11-13
GNP, 16

Government Communications Headquarters (GCHQ), 111
Growing complexity in international, 40-43
  first consolidation, or implementation, 40-43

H

High world trade growth vs output, 21-23
  globalisation, 21
  MFN clause, 22-23
  recent growth, 21
  trade growth, 21

I

ICT (Information and Communications Technologies), 129
IDA countries, 52
Illiteracy, 117
Inclusion or exclusion, 113-116
  analog-based infrastructure, 115
  communication networks, 114
  electronic commerce, 113
  ICT progress, 113
  income inequality and computers literacy gaps, 114-115
  information and communications technologies (ICT), 113
  information flows, 114
  leapfrogging stages of development, 115
  networked environment, 113
  technological determinism and government policies, 115-116
  welfare implications, 113
Indian Ocean, 109
Information and Communication Technologies (ICTs), 117, 118, 119, 120
Information and communication technology, 89
Information technology outsourcing goes global, 125-129
  call centres, 126
  Mauritius, 127
  Morocco, 127
  new a trend, 126-127
  North to South, or South to North, 127-129
  trade union, 128
  UK workers, 127
  United States, 128
International Chamber of Commerce, 107
– communications, 109
– migration, 60
– Monetary Fund, 4, 6, 7, 25, 47
  financial resources, 7
– telegram, 110
Internet, 100
Ivory towers, 121-124
  Australia, 122
  business, 121
  Canada, 122
  Cardean University, 122
    collaborative learning activities, 124
  doom and gloom scenarios, 124
  e-learning, 123, 124
  IT courses, 123
  jobs for life, 121
  online learning, 121
  Phoenix University, 123

race to partner up, 121
U.K., 122
UNext. Com, 122
United States, 122
University of Columbia (U.S.), 122

L

Labour pains, 106-108
battle to protect privacy, 106
clipper chip, 108
Communications Assistance for Law Enforcement Act, 107
cyber crime treaty, 106
cyber-rights, 106
Electronic Privacy Information Center, Human Rights Watch and the American Civil Liberties Association, 108
European Union (EU), 107
Global Internet Liberty Campaign (GILC), 108
gun-owners, 106
infamous clipper chip, 106
internet's growth, 107
old enemies become new allies, 107-108
Organisation for Economic Co-operation and Development (OECD), 107
privacy battle, 107
privacy-minded citizens, 106
Trans-Atlantic Consumer Dialogue (TACD), 108
U.S. congress, 108
war on terrorism, 107
Latin America, 35
Latin American countries, 33
Liberal conservative, 4
Liquidity arrangements, 7
Listening stations, 110
Listening, 109-112
fear not the world bomb, 110-111
spying on the government of the people, 111-112

M

Mandela, Nelson , 34
Market socialism, 10
Marrakesh Agreement, 86
Matter, 7
Measures of growth, 16
Migration, 59-63
Africa, 62
Asia, 62
costs and benefits, 61
global demographics, 59
Latin America, 62
policies, 62
policies, 62-63
question of gender, 61-62
refugees, 62
urban transformation, 60-61
Mike Moore, 88
Ministerial meeting in Marrakesh, 47
Mistaken arguments, 3
Mobile or m-commerce, 92
Modern communications, 6
Mostfavoured-nation (MFN), 22
Multilateral trade negotiations, 42
– trading system, 86
Mysore, 128

Myths and illusion, 38-39

N

NASSCOM (National Association of Software and Services Companies), 126
National and global population, 59
Nation-state and globalisation, 4-7
Negotiations on basic telecommunication, 42
Negotiations on maritime transport services, 43
Net gains or net dream, 117-120
techno babble, 119
Net visionaries, 118
New networks and services, 118
New Zealand, 110
NGO, 9, 25, 52, 54, 108
Nihilism, 2
Non-governmental organisations, 84
North American Free Trade Agreement (NAFTA), 15, 18

O

Obasajo, Olusegun, 88
OECD, 48

P

Peace and prosperity, 71-73
Afghanistan, 71
Angola, 71
Cambodia, 71
demise of the Austrian-Hungarian and Ottoman Empires, 72
Ethiopia, 71
first world war, 71
freedom and democracy, 72
Nicaragua, 71
Peru leftist government, 71
pluralism and democracy, 73
post colonial Africa, 72
religious nationalities, 72
Yugoslavia, 72
Poor health, 117
Poverty, 2, 11, 51, 117
– in the South, 64
– Reduction Strategy Paper (PRSP), 52
Privacy-friendly, 93
Privatisation, 4
Property right, 2

R

Radio, 110
Renewing the state, 24-25
Rich western European economic regions, 27

S

Safeguards, 23
Sanction system's sphere of influence, 29
Segregation, 9
Singapore issues, 86
South Africa, 24, 34
South Asia, 56
Southern firms export markets, 55-58
trade reform and macro-economic stability, 56
Supermarkets, 57

T

Technological changes, 15
Telecomes union, 125
Trade and investment, 3
Trade barriers, 57
Truth about global competition, 14-20

growth myth, 16-17
myth of free trade, 18-20
– – – unregulated markets, 17-18
– that corporations are benevolent institutions, 18-20
– – economic globalisation is inevitable, 18

U

U.S. Federal Trade Commission, 92
UK jobs to India, 125
UN Development Programme, 100
Undemocratic mafias, 9
Under-funded education, 117
United Nation Conference on Least Developed Countries in Brussels in May 2001, 83, 85
United Nations Development Programmes, 39
United Nations Millenium Declaration of September 2000, 83
United States, 6
Urban tensions, 9
Urbanisation and globalisation, 8-10
Uruguay round, 42-43, 46-47, 87
US job market, 125

V

Victory of capitalism, 2

W

West's policy markets, 27
World Bank, 4, 25
World intellectual property organisation, 116
World Trade Conference in Doha, 80
World Trade Organisation (WTO), 2, 15, 18, 25, 26, 28, 31, 41, 43, 46, 47, 80, 81, 82, 85, 86, 88, 48, 116
World trade, 21, 44-48
World trade: no losers in the round, 44-47
World trade: permanent negotiations, 47-48
World War II, 27, 37
World War III, 14
World's largest network, 111
Worsening environment conditions, 117